电池科普与环保②
Battery Science Popularization and Environmental Protection ②

电池简史

A Brief History of Batteries

（中英对照版）
(Chinese-English Version)

马建民 / 主编　　Edited by: Ma Jianmin
咪柯文化 / 绘　　Illustrated by: Micco Culture
李丝贝 / 译　　　Translated by: Li Sibei

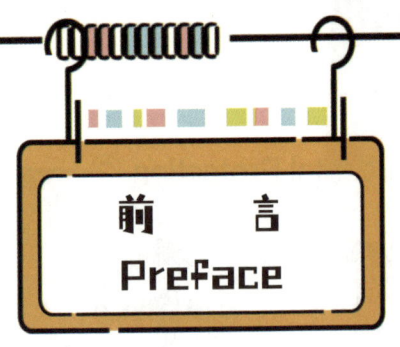

人类对能源的探索永不停止

人类对能源的探索，从来就未曾停止。由于地球上可供开采的煤炭、石油、天然气等非再生能源十分有限，因此，现在全世界都将目光聚焦于太阳能、风能、核能、潮汐能等再生能源的开发与利用。

能源问题是关系国家安全、社会稳定和经济发展的重大战略问题。优化资源配置，提高能源的有效利用率，对于人类的生存和国家的发展都具有十分重要的意义。

如何积极发展新能源是人类必须共同面对的一项重大技术课题。新能源技术的不断进步，特别是动力系统的不断改进，为能源结构的转型提供了可能。然而，虽然新能源的类型很多，但世界上至今还没有实用的、经济有效的、大规模的直接储能方式。因此，人类还不得不借助间接的储能方式。

电能，作为支撑人类现代文明的二次能源，它既能满足大量生产、集中管理、自动化控制和远距离输送的需求，又具有使用方便、洁净环保、经济高效的特点。因此，电能可以替代其他能源，提高能源的利用效率。

我们今天所有的可移动电子设备，其运行都离不开电池。电池的出现使人类的生活更加便捷，特别是在信息时代来临之后，电池的重要性更为突出。我国不仅是世界排名第一的电池生产大国，也是世界排名第一的电池消费大国。

人类虽然在电池的研究方面已经取得了丰硕的成果，但还一直在寻找更好的电能储存介质。随着科学的发展、新能源技术的成熟，在未来，哪一种类型的电池能够脱颖而出还未可知。希望此书能激发孩子们对电池的兴趣，让他们在未来为我们揭晓谜底。

2024 年 3 月

Endless Exploration of Energy

Since ancient times, humans' quest for energy has never ceased. Given the limited reserves of non-renewable energy like coal, oil, and natural gas on Earth, the use of renewable energy like solar, wind, nuclear, and tidal power, has become the new global focus.

Energy is a major strategic issue that bears on national security, social stability and economic development. How to allocate and use energy in a better way means a lot for both individuals and countries.

How to develop new energy is a major technological topic facing humanity. With the development of new energy technology, especially the power system, energy structure transformation is made possible. However, despite various kinds of new energy, there is yet to be a practical, cost-effective, large-scale way of direct energy storage. Therefore, we have to resort to indirect methods to store energy.

Electricity, as a secondary energy propelling modern civilization, can support mass production, centralized management, automated control, and long-distance transmission. At the same time, electricity is clean, economical, efficient, environmentally friendly and easy to use. We can replace other energy with electricity to use energy better.

All the electronic mobile devices today can't operate without batteries. Batteries make our life more convenient. Its importance grows even more prominent with the advent of the Information Age. China is now the world's biggest producer and consumer of batteries.

Although we have gained so much in battery research, researchers are still looking for a better medium for power storage. With progress in science and new energy technologies, which type of battery will stand out still awaits our exploration. Hopefully, this book will spur children's interest in batteries and one day make them tell us the answer in the future.

Ma Jianmin

March 2024

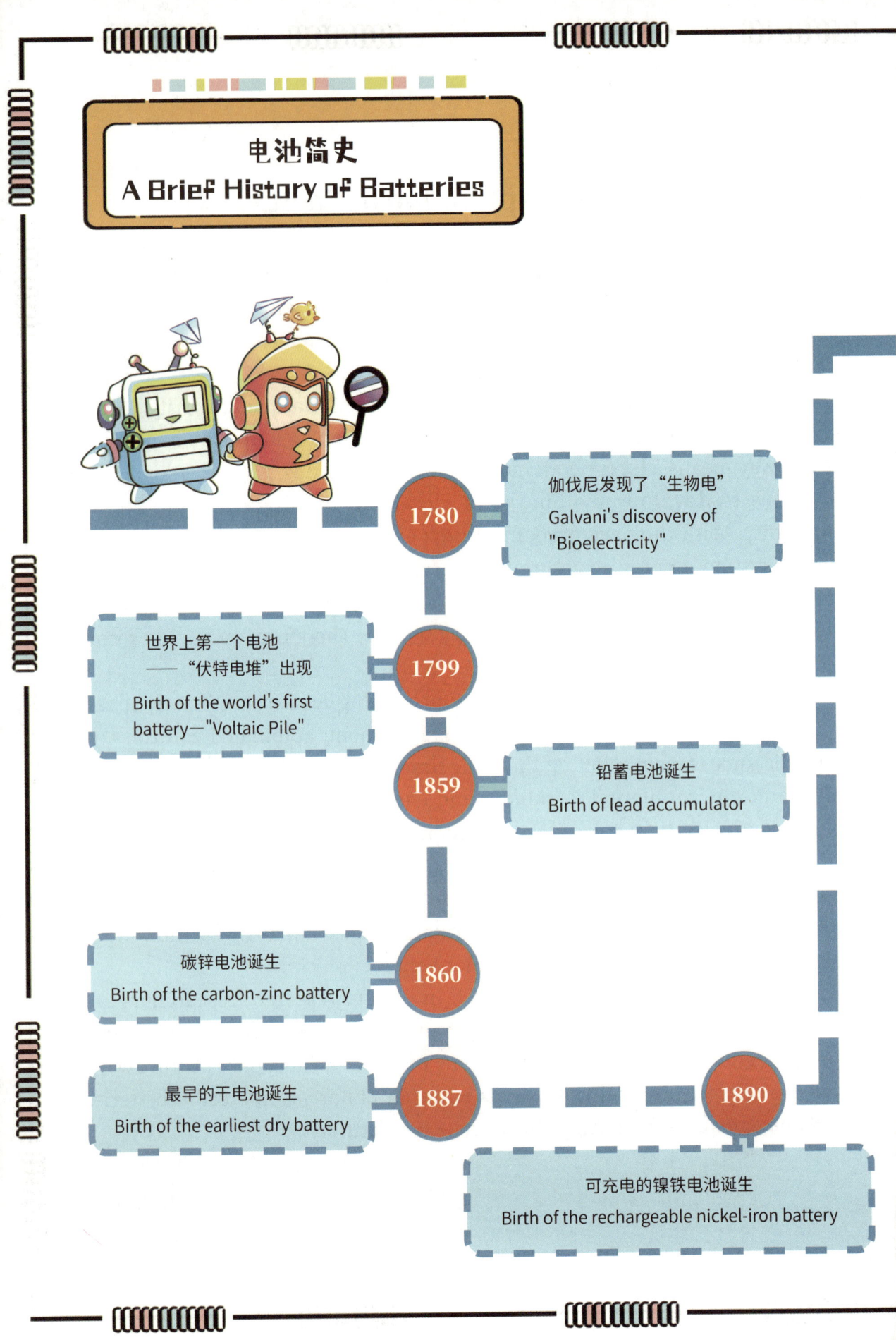

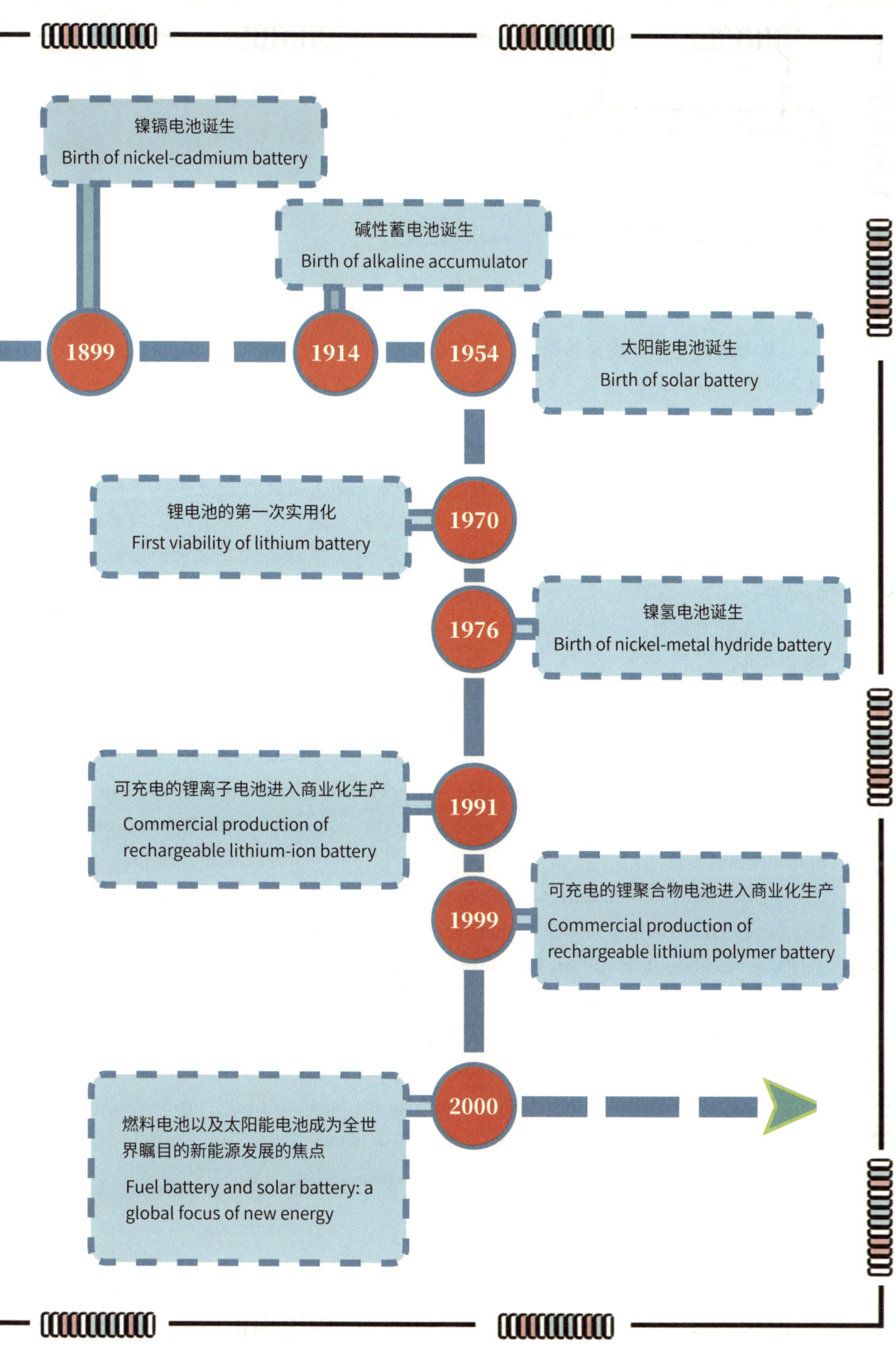

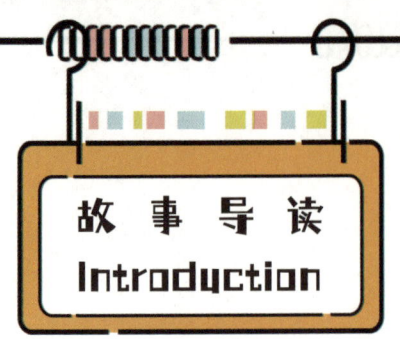

电池王国是一个庞大的国度，生活着许许多多的电池家族，每个家族的"电池人"都有着特殊的本领。他们勤劳能干，驱动各种设备运转，促进人类世界不断发展。

在电池王国，每天都有故事发生。在锂锂就任国王的仪式上，上一任国王王一硫送给了他一件宝物——一条精致的项链，这条项链将锂锂带入了一个奇异的空间。在那里，锂锂认识了居住在项链中的电池精灵——闪闪。闪闪告诉锂锂，他需要通过历史回廊来进行时光穿梭，完成帮助电池老前辈们解决烦恼的任务——集齐5颗宝石后才能成为真正的国王。

在历史回廊里，锂锂经历了哪些神奇的遭遇呢？他能顺利完成任务，集齐5颗宝石吗？

快一起来看看吧！

The Battery Kingdom is a huge country, with many battery families living there. Each family's "battery men" have special skills. They are hard-working and capable, driving all kinds of equipment and promoting the continuous development of the human world.

In the Battery Kingdom, there are stories every day. At the ceremony when Lithium Li takes the throne, the former king, King Monosulfide, gifts him a treasure— a delicate necklace. It transports Lithium Li to a magic world. There, he meets Shiny, a little spirit who lives inside the necklace. Shiny tells Lithium Li that he must get through the Historical Corridor to complete time-traveling missions, help battery ancestors solve their problems, and gather five gems to become a true king.

What magical adventures await Lithium Li in the Historical Corridor? Can he successfully complete his tasks and collect all the five gems?

Let's find out together!

角色介绍 Characters

锂锂　Lithium Li

家族：锂离子电池

Family: Lithium-ion Battery

大铅　Big Lead

家族：铅酸电池

Family: Lead-acid Battery

镍霸　Tyrant Nickel

家族：镍镉电池

Family: Nickel-cadmium Battery

机器人X　Robot X

王一硫　King Monosulfide

家族：钠-硫电池

Family: Sodium-sulfur Battery

闪闪　Shiny

身份：电池王国的守护精灵

Identity: Guardian Spirit of the Battery Kingdom

壹铅　First Lead
家族：铅酸蓄电池

Family: Lead-acid Battery

贰铅　Second Lead
家族：铅酸蓄电池

Family: Lead-acid Battery

布锌苦　Tireless Zinc
家族：锌锰干电池

Family: Zinc-maganese Dry

锌博士　Dr. Zinc
家族：锌-空气电池

Family: Zinc-air Battery

镍子楚　Child Nickel
家族：镍镉电池

Family: Nickel-cadmium Battery

目 录
Table of Contents

1 项链里的历史回廊
The Historical Corridor in the Necklace /001

2 大当家的烦恼
The Head's Worries /009

3 改造铅酸蓄电池
Remolding Lead-Acid Battery /017

4 宝石被盗了!
The Gem was Stolen! /029

5 遗失的心灵宝石
The Lost Mind Gem /037

电池大揭秘
Secrets behind Batteries
/053

电的发现
The Discovery of Electricity……………………………………054

电池王国起源之谜
The Mysterious Origin of the Battery Kingdom………………061

话说干电池
The Story of Dry Battery……………………………………066

话说蓄电池
The Story of Storage Battery………………………………070

话说太阳能电池
The Story of Solar Battery…………………………………077

话说锂电池
The Story of Lithium Battery………………………………087

话说燃料电池
The Story of Fuel Battery…………………………………104

话说碱锰电池
The Story of Alkaline Manganese Battery…………………115

话说镍镉电池
The Story of Nickel-cadmium Battery………………………117

未来电池之争
The Race among Future Batteries…………………………120

项链里的历史回廊
The Historical Corridor in the Necklace

瞧！电池王国的广场上礼炮齐鸣、烟花绽放，可以说是热闹非凡。

Look! In the square of the Battery Kingdom, cannons are fired in salutes and fireworks blossomed in the sky. How bustling!

电池王国国王就任仪式
The inauguration of the new king

新国王锂锂的就任仪式现在正式开始！

His Majesty, Lithium Li, now takes the throne!

请问国王，您的治国之策是什么呢？

Your Majesty, how will you govern our country?

前任国王曾说电池王国存在危机，这是真的吗？

The former king once said the Battery Kingdom was in crisis. Is that true?

广场上在干什么呢？原来是电池竞技大赛结束后，正在举行电池王国的新任国王——锂锂的国王就任仪式！

What's happening in the square? The inauguration ceremony for the new king, Lithium Li, is taking place right after the Battery Competition!

各位同胞，我非常荣幸能在国王这个岗位上为电池王国服务。我会兢兢业业、克己奉公，为王国的繁荣和发展作出贡献。

Fellow citizens, as your king, I am very honored to serve the Battery Kingdom. I will work diligently and selflessly to make our country prosperous.

我认为治国之道在于让各位同胞能够幸福地生活和工作，确保各个电池家族的发展和创新，为电池王国的未来蓄积能量，承前启后，让一代更比一代强！

I believe, the key to governing is to ensure the happiness and well-being of our people, and make sure every battery family can renew itself for greater progress. In that way, we'll gather steam for a brighter future and build on the past so that each generation becomes stronger than the last!

王一硫手里拿着的是什么？是传说中的国宝吗？

What is King Monosulfide holding in his hand? Is it the legendary national treasure?

在就任仪式上，前任国王王一硫将电池王国的传世国宝——电池项链，交给了锂锂。不过，这个项链看起来似乎并不完整，只有1颗蓝色宝石在闪闪发光，其余4颗宝石的位置还空缺着。

At the ceremony, the former king King Monosulfide passes on the heirloom of the Battery Kingdom—the Battery Necklace—to Lithium Li. However, the necklace seems incomplete, with only one blue gem shining brightly, and the locations of the remaining four gem being vacant.

尊敬的国王陛下，还请您多赐教治国良策！

Your Majesty, please share with us your wisdom on governing!

新的时代属于勇于创新、不断挑战自己的电池，由你来领导电池王国，我很放心。

The new era belongs to batteries that are good at innovation and challenge. I trust you to lead the Battery Kingdom.

现在，我将电池项链传承于你，你一定要好好保管，它关系到电池王国的命运。

Now, I pass the necklace on to you. You must keep it safe, for it holds the fate of our kingdom.

好的，我会铭记在心，不过，这个项链与您所说的危机有关系吗？

Okay, I will bear it in mind. But is this necklace connected to the crisis you've mentioned?

戴上它吧，项链会告诉你一切！

Put it on, and it will reveal everything to you!

典礼结束了。晚上，锂锂为自己接好电源就休息了。

突然，锂锂身上的电力竟然开始向脖子上的电池项链涌动，项链绽放出奇异的光彩，并逐渐放大，随后竟将锂锂吸入了另一个空间！

The ceremony is over. At night, Lithium Li plugs himself into a power source and then rests.

Suddenly, the power on Lithium Li begins to surge towards the battery necklace on his neck. The necklace blooms with a strange light and is gradually enlarged, and then Lithium Li is sucked into another space!

锂锂！快醒醒！
Lithium Li! Lithium Li! Wake up!

005

当锂锂再次睁开眼时，他正被身下的光环载着朝某个方向飞驰而去。他惊讶地看着周围的环境，全然不知发生了什么。

When Lithium Li opens his eyes again, he finds himself being carried in a certain direction by the halo under him. He looks at the surrounding environment in surprise, completely unaware of what is going on.

这是哪里？我的梦里吗？要往哪里去？
Where am I? Is this a dream? Where am I going?

终于，光环在一个发着光的物体面前停下来了，锂锂上前细细打量，发现这个发光物体竟然是一个正在沉睡的美丽精灵！

Finally, the halo stops in front of a glowing object. Lithium Li approaches to take a closer look and discovers that the shining thing turns out to be a beautiful spirit, slumbering in peace!

经过沟通，锂锂了解到，原来这是居住在电池项链里的精灵，名叫闪闪。当电池王国有新国王继任时，闪闪便会与之建立连接，协助新国王治理国家。

闪闪记录了每一种电池的起源与发展，能够带新国王进入历史回廊，进行时光穿梭。这也是每一任新国王继任后必须要完成的任务——从历史中学习宝贵的经验。每帮助历史回廊中的一种电池完成更新迭代，闪闪便会获得一颗宝石，集齐5颗宝石，闪闪才能拥有完整的力量，解锁项链的全部神奇功能。

After communicating, Lithium Li learns that this is an spirit named Shiny who lives inside the battery necklace. When a new king takes the throne, Shiny will establish a connection with him/her and assist in governing the kingdom.

Shiny has records of the origin and development of every type of battery. She can guide the new king through the Historical Corridor for time-traveling. This is a task that every new king must complete—to learn valuable lessons from history. Shiny can obtain a gem in helping a battery in the Historical Corridor with its updating and upgrading. Only by collecting all five gems can Shiny possess full power and unlock all the magical functions of the necklace.

这5颗宝石分别是时间宝石、空间宝石、力量宝石、灵魂宝石以及心灵宝石。

These five gems are: Time Gem, Space Gem, Power Gem, Soul Gem, and Mind Gem.

每一颗宝石的功能都不同，每获得一颗，我们就能解锁一个新功能！目前，我们只拥有时间宝石。

As every gem has its unique function, a new feature can be unlocked each time we acquire a gem! At present, we only own the Time Gem.

接下来，我将用它为你打开历史回廊，跟我来！

Then, I will use it to open the Historic Corridor for you. Follow me!

大当家的烦恼
The Head's Worries

闪闪带着锂锂停在了碱锰电池家族的历史片段面前，寻找宝石的旅途即将开始！

Shiny and Lithium Li stop in front of the historical segment of the alkaline-manganese battery family. The gem-finding journey is about to begin!

无论是哪一种文明的进步，都离不开一代接一代的传承。历史回廊记录了每一种电池的起源和发展，它们对电池王国的未来有着深远的影响。

The progress of any civilization relies on the inheritance from generation to generation. The Historic Corridoral records the origin and development of every type of battery, which has a profound impact on the future of the Battery Kingdom.

现在，我将带你进入第一个任务——为锌锰干电池家族的大当家解决烦恼！

Now, let's begin with your first mission—helping the head of the zinc-manganese dry battery family with his problems!

1868 年
乔治·勒克朗谢的实验室
1868, in Georges Leclanché's laboratory

什么？！这就开始了？
What?! Has it started already?

锌锰湿电池的诞生 The Birth of the Zinc-Manganese Wet Battery

闪闪发动魔法，和锂锂一起来到19世纪的法国工程师乔治·勒克朗谢的实验室，碱锰电池家族的祖先锌锰湿电池（也叫"碳锌干电池"）就是在这里诞生的。

Shiny casts a spell and travels with Lithium Li to Georges Leclanché lab in the 19th century. The ancestor of the alkaline-manganese battery family, the zinc-manganese wet battery (also known as "the carbon-zinc dry battery"), was born here.

碱锰电池家族之所以能在电池王国中有如此高的地位，很大一部分原因是他们历史悠久。

The reason why the alkaline manganese battery family has such a high status in the Battery Kingdom is mainly due to its long history.

至今，他们已经经历了100多年的升级迭代。喏，这就是他们诞生之初的模样。

Up to now, they have undergone more than 100 years of upgrades. Look, this is what they looked like at the very beginning.

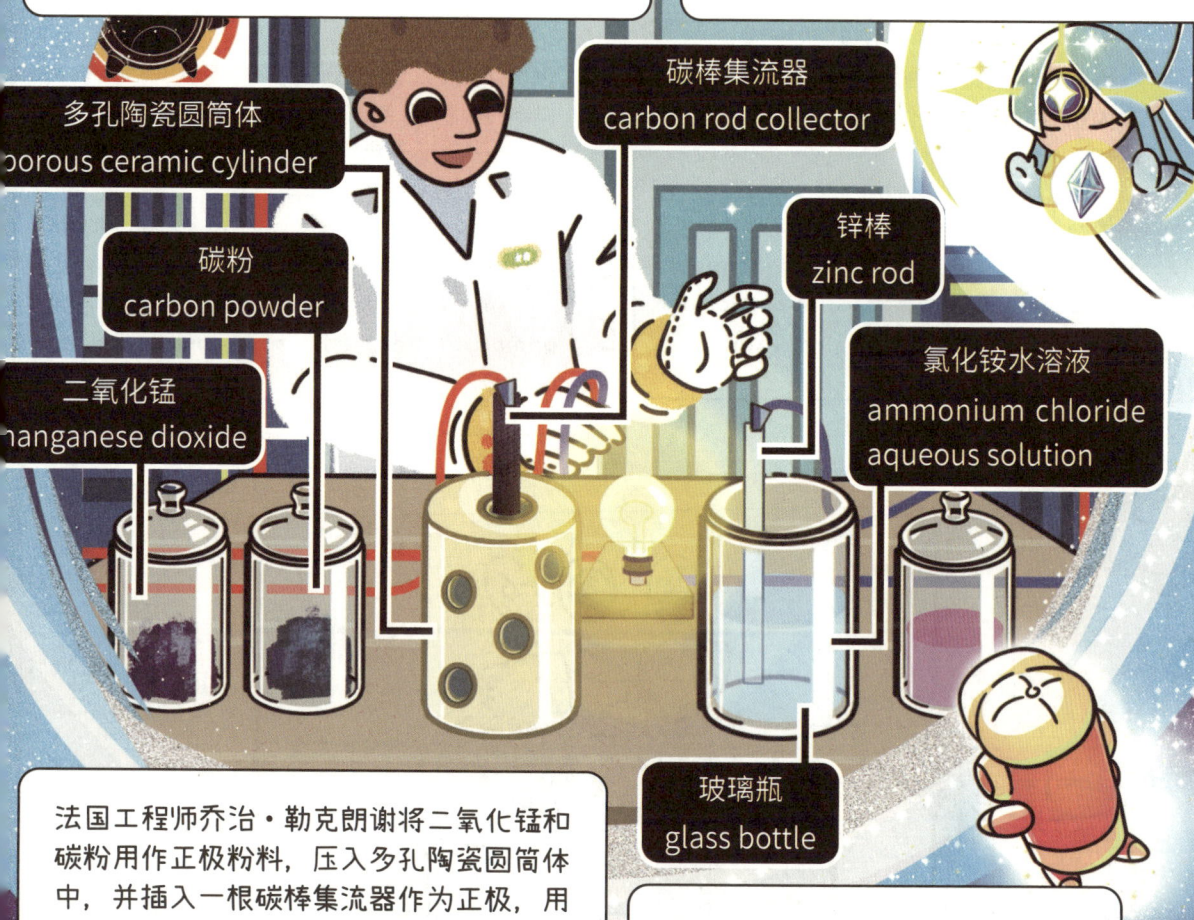

- 多孔陶瓷圆筒体 porous ceramic cylinder
- 碳粉 carbon powder
- 二氧化锰 manganese dioxide
- 碳棒集流器 carbon rod collector
- 锌棒 zinc rod
- 氯化铵水溶液 ammonium chloride aqueous solution
- 玻璃瓶 glass bottle

法国工程师乔治·勒克朗谢将二氧化锰和碳粉用作正极粉料，压入多孔陶瓷圆筒体中，并插入一根碳棒集流器作为正极，用一根锌棒部分插入溶液中作为负极。

Georges Leclanché used manganese dioxide and carbon powder as the positive electrode materials, pressing them into a porous ceramic cylinder. Then, he inserted a carbon rod collector as the positive electrode and had a zinc rod partially immersed in the solution as the negative electrode.

电池的电解液是浓度为20%的氯化铵水溶液，电池的容器是玻璃瓶，就这样他制成了第一个锌锰湿电池。

The electrolyte of the battery is a 20% ammonium chloride aqueous solution, and the container is a glass bottle. This is how the first zinc-manganese wet battery was made.

1887 年，干电池的诞生　1887: The Birth of the First Dry Battery

参观完锌锰湿电池的诞生，闪闪又带锂锂见证了碱锰电池发展历史上的第二个重要节点——锌锰干电池的诞生。

1887 年，丹麦人威廉·赫勒森在最初的碳锌电池的基础上进行改进，制出了原电池雏形，即锌锰干电池。

After exploring the origin of the zinc-manganese wet battery, Shiny takes Lithium Li to witness another key moment in the history of alkaline-manganese battery—the birth of the zinc-manganese dry battery.

In 1887, a Dane, Wilhelm Hellesen improved upon the early carbon-zinc battery and created the first prototype of the primary battery, known as the zinc-manganese dry battery.

赫勒森将碳锌电池的电解液——氯化铵水溶液，改为由氯化铵、氯化锌、石膏和水组成的糊状物。

Hellesen replaced the carbon-zinc battery's electrolyte—ammonium chloride aqueous solution—with a paste composed of ammonium chloride, zinc chloride, gypsum, and water.

他又将锌片做成圆筒作为电池的容器，同时用石蜡封口，制出了原电池的雏形——锌锰干电池。

He shaped a zinc sheet into a cylindrical container for the battery and sealed it with paraffin wax, creating the prototype of the primary battery—the zinc-manganese dry battery.

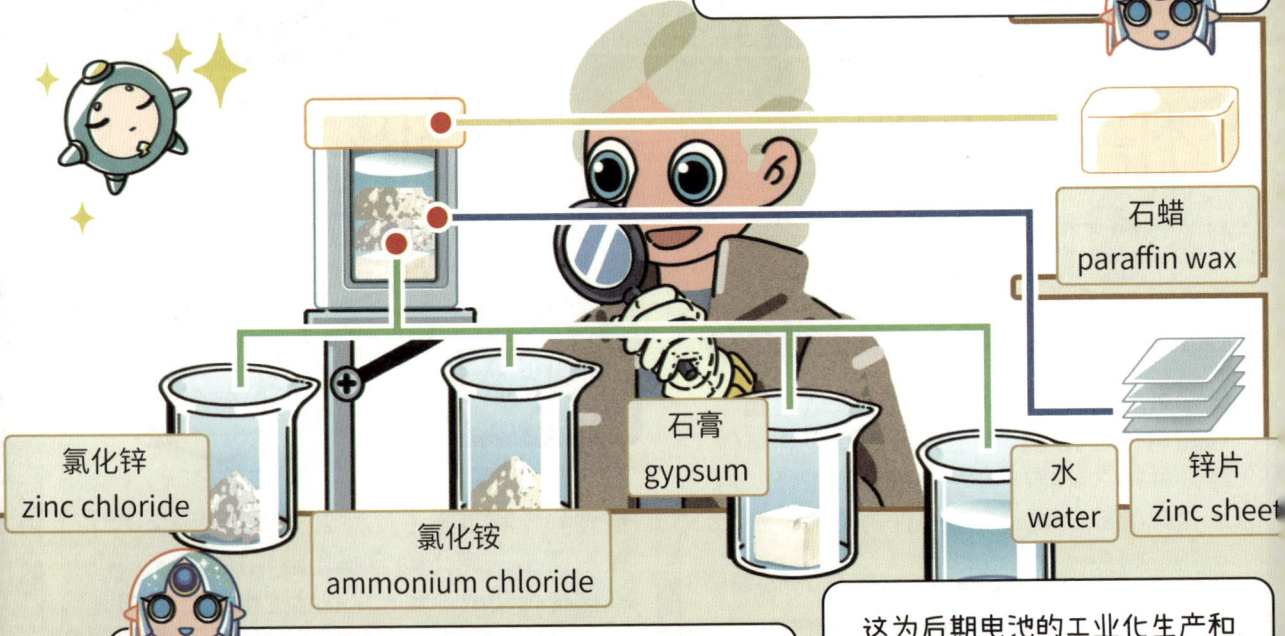

氯化锌 zinc chloride
氯化铵 ammonium chloride
石膏 gypsum
水 water
锌片 zinc sheet
石蜡 paraffin wax

此后不久，人们在这个基础上又将面粉和淀粉作为电解质溶液的凝胶剂，使电池的便携性大大提高。

Soon after, flour and starch were used as gels for the electrolyte, making the batteries much more portable.

这为后期电池的工业化生产和广泛使用打下了良好的基础。

This laid a good foundation for the industrial production and wide use of later batteries.

1890: The Industrial Production of Zinc-manganese Dry Batteries

锌锰干电池自从诞生以来，就在不断地升级迭代，其性能不断提高。从1890年开始，锌锰干电池就已经在全世界范围内工业化生产。

在工厂的流水线上，锂锂和闪闪见到了那个时候的锌锰干电池家族的大当家。大当家向锂锂寻求帮助，这应该就是此次的任务了！锂锂痛快地答应下来。

The zinc-manganese dry batteries have been continuously upgraded and iterated with their better performances ever since their birth. From 1890 on, they have been in industrial production worldwide.

On the assembly line, Lithium Li and Shiny meet the head of the zinc-manganese dry battery family. The head seeks help from Lithium Li. This should be the task! Lithium Li says yes with a big smile.

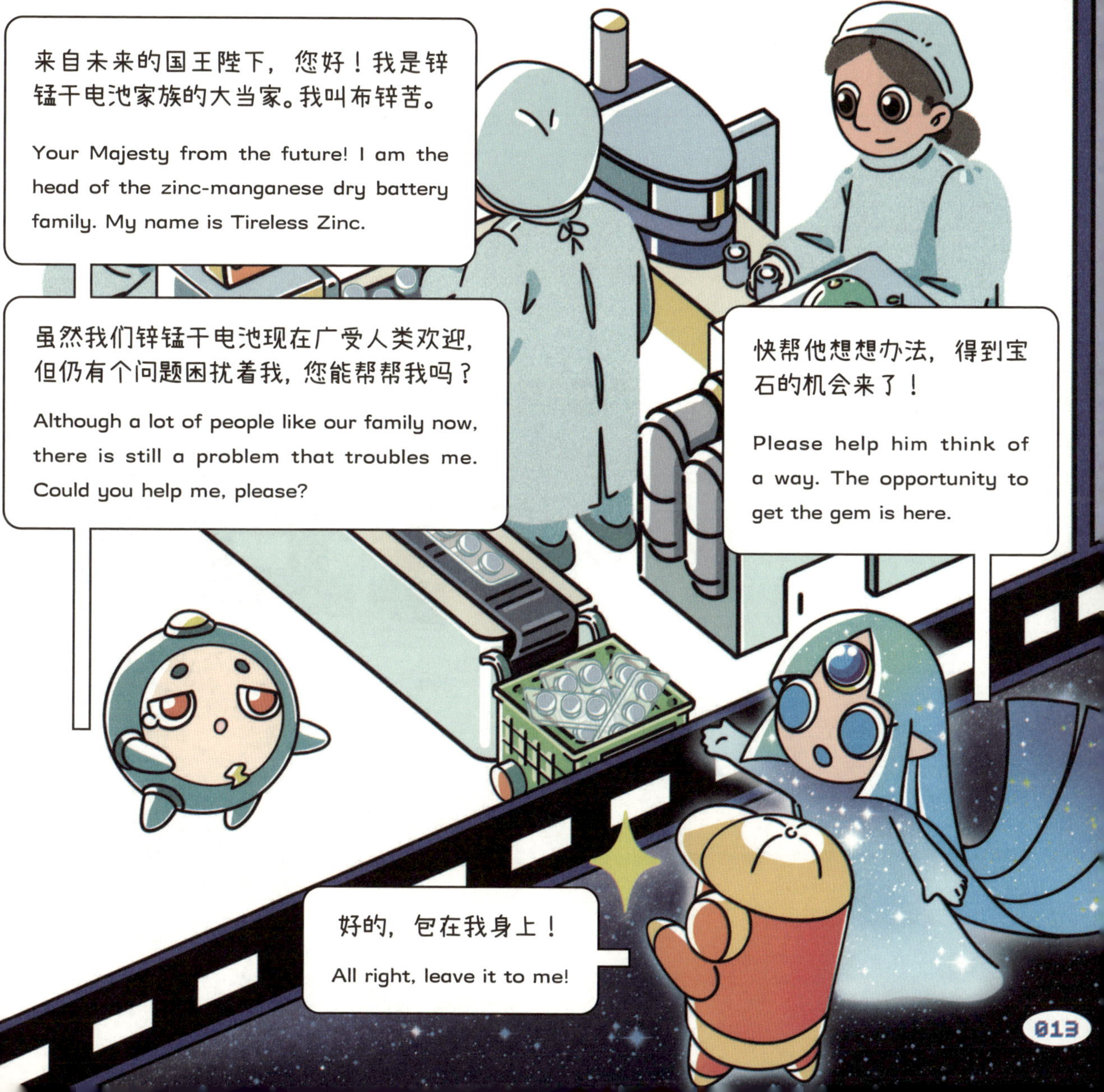

> 来自未来的国王陛下，您好！我是锌锰干电池家族的大当家。我叫布锌苦。
>
> Your Majesty from the future! I am the head of the zinc-manganese dry battery family. My name is Tireless Zinc.

> 虽然我们锌锰干电池现在广受人类欢迎，但仍有个问题困扰着我，您能帮帮我吗？
>
> Although a lot of people like our family now, there is still a problem that troubles me. Could you help me, please?

> 快帮他想想办法，得到宝石的机会来了！
>
> Please help him think of a way. The opportunity to get the gem is here.

> 好的，包在我身上！
>
> All right, leave it to me!

原来，虽然那时的锌锰干电池家族风头正盛，但他们的大当家已经意识到，锌锰干电池自身的性能不够有竞争力，因而对自家的未来感到担忧。锂锂很快就想到了解决办法，那就是向他介绍锌锰干电池家族在21世纪的发展情况，希望能让他放下心来。

It turns out that although the zinc-manganese dry battery family is in the limelight, the head knows that they are not competitive enough. That is why he is worried about the future. Lithium Li quickly comes up with a solution. He introduces the development of the zinc-manganese dry battery family in the 21st century to him, hoping to reassure him.

随着电气时代的来临，电力成为人类生产与生活的主要能源，眼见各种电池"新秀"不断崛起，当下的我们却有很大的缺陷。

With the advent of the electric age, electricity has become the main energy source for human production and life. Seeing the rise of various battery "rookies", we have great flaws today.

- 绝缘物质 insulating material
- 二氧化锰 + 碳粉 manganese dioxide + carbon powder
- 石墨 graphite
- 二氯化锌 + 氯化铵 zinc dichloride + ammonium chloride
- 锌 zinc

锌锰干电池（酸性锌锰电池）
zinc-manganese dry battery (acid zinc-manganese battery)

我们在工作的过程中，电压会持续下降，不能提供稳定的电压，而且还有放电功率低、比能量小、低温性能差等问题。

During our work, for instance, the voltage will continue to drop, unable to provide a stable voltage. Other problems include low discharge power, small specific energy, and poor low-temperature performance, etc.

如果不居安思危，我担心我们很快就被淘汰了……
If we don't plan ahead, I'm afraid we'll be replaced soon…

别担心！你们将会经历很多次升级，即使到了21世纪，你们仍然是干电池中产量最大的电池！

Don't worry! You will experience many upgrades, and even in the 21st century, you will still be the type of battery with the highest production volume in dry batteries!

特别是1912年以后，你担心的问题都会被人类解决。而且，你们家族会诞生一个非常杰出的后辈——碱性锌锰电池。

Especially after 1912, all your worries will be solved by humans. And guess what? Your family will have a super awesome new member—the alkaline zinc-manganese battery.

碱性锌锰电池是由你们锌锰干电池升级而来，可以说是最优秀的高容量干电池之一。

The alkaline zinc-manganese battery is an upgrade from your zinc-manganese dry battery and can be said to be one of the most successful high-capacity dry batteries.

看！他们的结构与你们相反，负极在内、正极在外，人们称之为"反极结构"，这样更适合大电流连续放电。

Look! Their structure is the opposite of yours. The negative pole is inside while the positive one is outside. People call their structure "the inverse one", which allows for continuous high-current discharge.

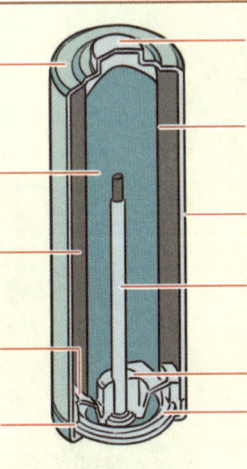

- steel shell　钢壳
- negative electrode material　负极物料
- positive electrode material　正极物料
- seal ring　密封环
- sealant　密封剂
- 正极帽　positive terminal cap
- 隔离膜　isolating film
- 导电膜　conductive film
- 集电铜针　copper collector pin
- 密封盖　sealing cover
- 负极盘　negative plate

碱性锌锰电池
alkaline zinc-manganese battery

并且，电池中原材料的纯度和用量都提升了，更是将你们现在的电解液由盐换成了强碱。这样一来，他们不仅电容量大，还具备优良的低温性能、储存性能和防漏性能。

Besides, the purity and dosage of raw materials in the battery have been improved, and your current electrolyte has been replaced by a strong alkali. In this way, they not only have high capacity, but also have excellent low-temperature performance, storage performance, and leak-proof performance.

短短几十年，竟发生了如此大的变化！人类的智慧真了不起！

In just a few decades, such great changes have taken place! Human wisdom is truly amazing!

另外，随着人们环境保护的意识的增强，碱性锌锰电池更是实现了低汞化和无汞化！在未来……

Moreover, as people care more about our environment, these batteries have become low-mercury and even mercury-free! In the future…

哈哈哈！如此精妙的设计，真是"长江后浪推前浪"啊！

Hahaha! What a brilliant design! It shows that "each generation surpasses the one before it"!

015

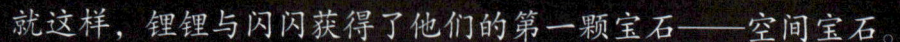

改造铅酸蓄电池
Remolding Lead-Acid Battery

大铅收到锂锂被不知名光环"绑架"的消息后，四处搜寻却不见锂锂的踪影。当他正在房间里焦急地来回踱步时，锂锂又被光环完好地送回来了。

　　锂锂刚一落地，大铅就飞扑到锂锂面前仔细询问起来，锂锂便向他讲述了这一夜的经历。

　　大铅惊讶地听完锂锂的讲述，激动不已，他迫不及待地也想去这个神奇的空间一探究竟！

On hearing that Lithium Li has been "kidnapped" by a mysterious halo, Big Lead searches everywhere but finds no trace of Lithium Li. As he is anxiously pacing back and forth in his room, the halo safely returns Lithium Li.

As soon as Lithium Li lands, Big Lead hurries to him and asks carefully about what happened. Lithium Li tells him about the experience of that night.

Big Lead feels excited as he listens to Lithium Li's story. He couldn't wait to go to this magical space to find out!

没想到这么轻松就完成了任务，然后我就被送回来了。

I can't believe the mission was completed so easily, and then I was sent back.

历史回廊？住在项链里的精灵？还可以穿越时空？太神奇了！

The Historical Corridor? An spirit living inside a necklace? And time travel too? It's incredible!

锂锂和大铅紧张地向项链输送电力,但没有出现想象中的异空间,闪闪倒出现了。

一见到闪闪,大铅便开始了连环发问,闪闪耐心地一个个解答着。

Lithium Li and Big Lead are straining to deliver electricity to the necklace. However, no supernatural space appears as they imagined. Instead, here comes Shiny.

On seeing Shiny, Big Lead starts firing questions at her. Shiny patiently answers one by one.

你……你就是住在项链里的精灵?

Are… are you the spirit living inside the necklace?

没错!我就是大名鼎鼎的电池精灵——闪闪!

Yes! My name is Shiny, the famous battery spirit!

你们的下一次任务是什么呢?

What is your next mission?

下一次任务是——拯救颓废的退役铅酸蓄电池。

The next mission is—save the run-down retired lead-acid battery.

你在项链里面都做些什么呢?

What do you do inside the necklace?

项链里可无聊了,大多数时间我都在睡觉!

Oh, it's quite boring. Most of the time, I'm sleeping!

哇!你还能到外面的世界来?

Wow! You can come out into the outside?

那当然!有了时间宝石和空间宝石,我可以去到任何时间任何地点!

Of course! With the Time Gem and Space Gem, I can go anywhere, anytime!

听到下一次任务与自己的家族有关，大铅瞪大了双眼，嚷嚷着一定要一起去。闪闪随即开始发动魔法，新的旅程即将开始。

谁都没有注意到的是，角落里的镍霸正偷偷地观察着这一切。

Hearing that the next mission is related to his family, Big Lead widens his eyes and yells that he must go together. Shiny then cast her magic, and a new journey is about to begin.

However, no one notices that Tyrant Nickel is secretly observing all this in the corner.

我们家还有颓废的电池？我也想一起去看看！

We have run-down batteries? I want to go and see as well!

闪闪，我可以带着我的朋友大铅，一起进入历史回廊吗？

Shiny, may I bring my friend Big Lead with me into the Historical Corridor?

当然可以！咱们这就出发！

Of course you can! Let's go now!

时空隧道？这里面一定有宝藏！

A time-space tunnel? There must be treasures inside!

转眼间，三人来到了1859年的法国。在一间实验室里，他们见到了刚刚诞生的铅酸蓄电池以及铅酸蓄电池的发明者——加斯东·普朗泰。

In a flash, the three of them arrive in France in the year 1859. In a laboratory, they see the newly-born lead-acid battery and Gaston Planté, the inventor.

呜呜呜，老祖宗，见到您真是太激动了！没有您，就没有我们家族的今天！

Wow, my great…great-grandfather! It's so exciting to meet you! Without you, our family wouldn't have existed today!

你们好，来自未来的朋友们！

Hello, friends from the future!

自从普朗泰在1859年发明了以稀硫酸为电解质的第一块铅酸电池，铅酸电池的发展就正式揭开了序幕。

Since Planté invented the first lead-acid battery with dilute sulfuric acid as the electrolyte in 1859, the development of lead-acid batteries officially began.

没等大铅与他的老祖宗叙旧，闪闪又将时空切换到了21世纪的人类世界。在热闹的大街上，他们看到人类驾驶着由铅酸电池驱动的电动车来来往往。

Before Big Lead has a chance to talk with his ancestor, Shiny switches the spacetime to the 21st century.

On the bustling streets, they see people driving electric vehicles powered by lead-acid batteries to and fro.

当普朗泰在100多年前发明铅酸蓄电池时，他可能无法想象，他的此项发明有一天能占据世界上二次电池市场的最大份额。

When Planté invented the lead-acid battery over 100 years ago, he could hardly imagine that his invention would one day dominate the world's secondary battery market.

铅酸蓄电池由于其材料廉价、工艺简单、技术成熟、自放电低、免维护等特性，被广泛应用于不间断电源、电网和汽车等领域。

Lead-acid batteries are widely used in uninterruptible power supply, power grid and automobile due to their characteristics of cheap material, simple process, mature technology, low self-discharge and maintenance-free.

大铅正骄傲着,画面又一转,他们来到了一个与刚才的热闹情景截然相反的地方。

锂锂和大铅被眼前的景象震惊了。在这里,废弃的铅酸蓄电池堆积成山,退役的铅酸蓄电池横七竖八、死气沉沉地躺在街头、坐在地上,活脱脱一副"难民营"的景象。

Big Lead is beaming with pride when, with a sudden shift, they arrive at a place starkly contrasting to the bustling scene they just left.

Lithium Li and Big Lead are shocked by the sight. Here, discarded lead-acid batteries are piled up like mountains, and those used-up ones are strewn about, lying lifelessly on the streets and sidewalks. What a scene of "refugee camp"!

在如此庞大的使用规模之下,每年退役的铅酸蓄电池约有数百万吨,并且还在逐年增长。

With such a massive scale of use, millions of tons of lead-acid batteries retire each year, and this number is growing each year.

铅酸蓄电池的主要原料是铅,铅作为人类世界公认的有毒重金属,对人的身体以及土壤环境都具有极大的破坏力,如果不妥善处理,将会造成非常严重的后果。

Lead, as the main ingredient of lead-acid batteries, is a well-known toxic heavy metal. It poses a great threat to human health and the soil environment, and if not properly handled, it can cause severe consequences.

大铅飞快地跑到距离他最近的族人面前，已经退役的铅酸电池——贰铅，开始诉说他们的处境。

　　Big Lead dashes to a clan man nearest to him. Second Lead, an old and retired lead-acid battery, starts telling what happened to them.

集中处理站里的我们之所以如此颓废，并不是因为惧怕寿命耗尽。

We feel depressed in this centralized processing station, not because we are afraid of running out of life.

当我们退役之后，或是寿命将尽的时候，就会被人类收集起来，统一放在一处。

We are due to be collected here after our retirement or near the end of our life.

在我们的有生之年，我们都尽可能地发过光、发过热，为人类作出过贡献，我们为此感到非常骄傲。

In our lifetime, we've done our best to shine and heat up to help people. We're very proud of that.

原来，让废旧铅酸电池心情低沉的原因有两点：一是为彻底报废之后的遭遇而感到恐惧，二是为家族的未来而感到担忧。

锂锂决定，带着贰铅一起去未来体验一下新技术，让他知道，人类并没有放弃他们，而是找到了变废为宝的办法。

The retired lead-acid batteries are sad for two reasons: they are scared of their fate once being completely discarded; and they also worried about the future of their family.

Lithium Li decides to take Second Lead to the future to experience new technologies and show him that humans have never given up on them. Instead, they have found ways to turn old batteries into something useful again.

我知道了！21世纪的科学家们已经研究出了废旧铅酸电池无损修复技术！

I know! Scientists in the 21st century have developed a non-destructive repair technology for waste lead-acid batteries!

闪闪，你有办法带着他一起穿越时空吗？

Shiny, can you help us travel through time?

没问题，看我的！

No problem! Leave it to me!

闪闪将三人一起送到了铅酸电池无损修复技术已经成熟的年代。借助电子脉冲扫频振荡技术，贰铅重获新生。

Shiny takes the three them to an era when the technology for non-destructive repair of lead-acid batteries has fully developed. With the help of electronic pulse sweep-frequency oscillation technology, Second Lead is brought back to life.

即使在某一个阶段的发展中遇到了瓶颈，你们也不要感到悲观，人类是不会轻易放弃任何一种电池的！即使你们的体内含有有害物质，人类也会积极想办法处理。

Even if you encounter a bottleneck in a certain stage of development, don't feel pessimistic. Humans will not give up any kind of battery easily! Even if your body contains harmful substances, humans will actively find ways to deal with them.

例如，你们是所有电池中回收率最高的，所以你们在退役后，会去到集中处理站。经过人类有效的控制措施，一般情况下，你们并不会污染环境。

For example, you have the highest recycling rate among all batteries, so after you retire, you will go to a centralized processing station. After effective human control measures, under normal circumstances, you will not pollute the environment.

哇，这就是宝石吗？这一路我太激动了，都忘记了任务的事儿！

Wow, is that the gem? I'm so excited all the way that I almost forget my mission!

经过修复之后，贰铅无论是身体还是心理都重新回到了健康的状态，锂锂成功完成了第二个任务，力量宝石顺利归位。

After being repaired, Second Lead becomes healthy and strong again, both in body and spirit. With Lithium Li's completion of the second task, the Power Gem is restored to its rightful place.

> 感谢你们的帮助，我回去之后，要把你们带给我的力量同样传递给我的族人们，让他们也振作起来！
>
> Thank you for your help. When I go back, I will share with my people the same strength you have given me and cheer them up!

> 对未来充满希望，家族才会有力量！作为感谢，我将这颗力量宝石送给你们！
>
> Hope for the future gives our family strength! To show my thanks, I want to give you this Power Gem!

4

宝石被盗了!

The Gem was Stolen!

自从锂锂继任以来，电池王国一片欣欣向荣。锂锂的声望日益高涨，深受电池人们的爱戴。

Ever since Lithium Li became the new king, the Battery Kingdom has been full of energy and happiness. Everyone loves Lithium Li, and his popularity keeps growing every day!

> 我们电池王国真是发展得越来越好了！
> Our Battery Kingdom is getting better and better!

电池王国 上天入地 大显神通！
The Battery Kingdom, soaring high and diving low, shows off its magic glow!

自从锂锂当上国王之后，带动了不少电子产品的发展，电池王国的各个家族在人类世界上天入地，大显神通！

Since Lithium Li became the king, he has spurred the development of many electronic products! Thanks to him, the battery families have been everywhere in the human world, showing off their magic glow!

这天，锂锂与大铅正准备召唤闪闪进行下一次任务时，项链突然发出一阵异样的红光，紧接着，闪闪一脸慌张地出现了。

One day, Lithium Li and Big Lead are about to summon Shiny for the next mission. The necklace suddenly gives off a strange red glow. There appears Shiny in panic.

闪闪？！
Shiny?!

你怎么了？
What's wrong?

历史回廊出事了！
快跟我来！
The Historical Corridor is in trouble! Quickly, follow me!

锂锂和大铅赶紧跟随闪闪进入了空间，来到了历史回廊，对每一个回忆片段进行排查。在任务记录里，锂锂看到了一个熟悉的身影——镍霸。

Lithium Li and Big Lead quickly follow Shiny into the space and arrive at the Historic Corridor. Together, they begin checking each memory segment. In the mission log, Lithium Li spots a familiar figure—Tyrant Nickel!

你们看到的红光，是项链的空间系统发出的警报信号。

The red light you see is an alert from the necklace's space system.

这是在提示空间内发生了异常情况，需要尽快找到缘由并且处理，否则，有可能导致系统错乱，后果将非常严重！

It's warning us that something unusual is happening in the space! We need to find out what's wrong and deal with it quickly, or the system might go out of control, and that could lead to big trouble!

到底是哪里出了问题？

What's wrong?

会不会是上一次的任务……我们遗漏了什么？

Maybe it's the last mission… What did we miss?

快看！镍霸怎么会在这里？

Look! How can Tyrant Nickel be here?

三人赶紧依据记录中镍霸曾出现的场景，回到了之前那个废弃铅酸电池集中处理站。锂锂本以为，完成上一次任务之后，这里的电池们应该能够振作起来，没想到的是，他们看起来更加萎靡了。他们神情冷漠、目光呆滞，口中似乎还在念念有词地说些什么。

锂锂找到了贰铅，发现他时，他正躲在一个垃圾桶里面瑟瑟发抖。

The three of them quickly follow the clues in the records and return to the discarded lead-acid battery centralized processing station where Tyrant Nickel had once appeared. Lithium Li thought the batteries would be in better condition after their last mission, but to his surprise, they look even worse! Their expressions are cold, their eyes dull, and they are mumbling something strange under their breath.

Lithium Li finds Second Lead shivering inside a trash bin.

前辈，你怎么会在这里？
Sir, why are you here?

发生了什么事？别怕，我们来帮助你们了。
What happened? Don't be afraid, we're here to help you.

通过贰铅，三人了解到，镍霸竟用心灵宝石控制了众多电池人，才导致这些电池人性情大变，在这背后，似乎还隐藏着什么更大的阴谋。

Second Lead tells them that Tyrant Nickel used the Mind Gem to control many battery men, causing them to act in strange ways. There seems an even bigger scheme behind it.

自从你们离开后，有一个与你们来自同一时期的镍镉电池家族的电池人，使用心灵宝石，蛊惑了我的族人们，让他们变得易怒、好斗，并且仇视人类，甚至还憎恨自己的家族与王国。

After you left, a battery man from the nickel-cadmium family—someone from your time—came here. He activated the Mind Gem and bewitched my people, making them angry and aggressive. They hate humans and even hate their own family and kingdom.

这个坏蛋还用同样的方式控制了一大帮电池，照这样下去，我们整个世界都要乱套了！

That bad guy also used the same trick to control a group of other batteries. If this continues, our whole world will be in chaos!

什么？！镍霸拥有心灵宝石？！

What?! Tyrant Nickel has the Mind Gem?!

我猜测，应该是上一次我带你们进行时光穿梭时，镍霸也在场，于是就被一同带入了这个时期。

I guess that Tyrant Nickel was also present the last time when I took you through time.

拥有心灵宝石，就能拥有掌控人心的力量。心灵宝石一旦落入心怀不轨的人手上，后果将不堪设想！他应该还在这儿，我们快去找找看！

The Mind Gem has the power to control people's mind. Once the Mind Gem falls into the hands of someone with bad intentions, the consequences will be unimaginable! He should still be here. Let's go and find him.

殊不知，三人刚一回到这个时空，就被镍霸盯上了。"仇人"见面，分外眼红！镍霸迫不及待地出现在几人面前，命令手下的电池人将他们团团包围起来。

Little have they known they are targeted by Nickel Tyrant the moment they return to this time and space. When enemies meet, their eyes blaze with hatred! Nickel Tyrant can't wait to appear in front of them and orders his battery men to surround them.

> 不用你们找！本王就在这儿！
>
> You don't need to look for me—I'm right here!

> 哈哈哈！让本王来教教你，什么样的国王才叫威风！
>
> Hahaha! Let me show you what a true king looks like!

> 闪闪！快带我们离开这里，去镍镉电池家族被发明的时空！
>
> Shiny! Get us out of here quickly! Take us to the era when the nickel-cadmium battery family was invented!

5

遗失的心灵宝石
The Lost Mind Gem

眼见无路可走，锂锂当机立断，令闪闪打开时空之门，带领众人来到了1899年镍镉电池诞生的实验室。

Seeing no way out, Lithium Li made a quick decision. He turned to Shiny and said: open the time-and-space door! Together, they came to the laboratory where the nickel-cadmium battery was born in 1899.

希望能从源头上找到解决办法……
We hope we can find a solution from the source…

他有宝石，我们也有！这就叫——用魔法打败魔法！
He has a gem? Well, so do we! This is what they call—fight magic with magic!

一定要想办法把宝石拿回来！
We must find a way to get the gem back!

呼呼，太可怕了！
Whoo-hoo, so scary!

火灾报警器　Fire Alarm

作为一种为人类火灾警报系统提供稳定电源而研发的电池，初代镍镉电池——镍子楚，能够通过自己的能力来保护人类的安全，他为此感到非常骄傲。

听完众电池的经历之后，镍子楚为他的后辈伤害同胞的行为感到愤慨不已，当即表示要加入阻止镍霸的队伍。

The first-generation nickel-cadmium battery, Child Nickel, was created to provide a stable power source for the fire alarm systems. He felt proud and happy knowing his energy can help keep people safe.

After hearing what had happened to the other batteries, Child Nickel was shocked and angry. He couldn't believe his descendants were hurting others. Without hesitation, he stood up and said he wanted to help stop Tyrant Nickel.

……于是，我们就逃到了你这里。

…So we escaped here.

他竟惹出了这么大的祸端！作为镍镉电池家族的始祖，我难辞其咎。

He has caused such a huge mess! As the ancestor of the nickel-cadmium battery family, I can hardly obsolve myself from the blame.

后续的行动请让我加入，希望能助你们一臂之力！

Please let me join your mission. I hope to lend you a hand!

在闪闪的帮助下,一行人来到了 20 世纪 50 年代。

在一处军事基地内,镍镉电池士兵们正热火朝天地训练着,坦克里、飞机上,随处可见他们的身影。

With Shiny's help, the group came to the 1950s.

At a military base, nickel-cadmium battery soldiers were training hard, and they could be seen everywhere in tanks and on airplanes.

难以置信,我们镍镉电池家族居然能取得这样的成就!真为他们感到骄傲!

I can't believe our nickel-cadmium battery family has achieved this! I'm so proud of them!

或许……我们能从这里获取一些武器与装备?

Perhaps… we can get some weapons and equipment here?

在你被研制出来之后,经历了几次重要的改进。

After you were developed, there were several important improvements.

1932 年,人类科学家在你的体内加入了一种活性物质,进一步增强了这一代镍镉电池的导电性。

In 1932, human scientists added an active substance to you, further enhancing the conductivity of this generation of nickel-cadmium batteries.

嘘!

Shh!

全是镍镉电池,好可怕……

There are all nickel-cadmium batteries. It's horrible…

为了增强团队战斗力，趁着军队休息的间隙，锂锂带着大家来到一间装满武器的仓库前，打算挑选几样趁手的武器。

不料，他们被几名留守的士兵发现了。看到与周围环境格格不入的几人，士兵们举起手中的枪就准备攻击。

To enhance the team's combat power, Lithium Li led everyone to a warehouse full of weapons while the army was taking a break. They wanted to find some good weapons.

Unexpectedly, they were discovered by several guarding soldiers. Seeing them out of tune with the surrounding environment, the soldiers raised their guns and prepared to attack.

不要动！举起手来！
Don't move! Hands up!

你们从哪里来？要干什么？
Where are you from, and what are you doing here?

在镍子楚解释了事情原委后，这几名士兵决定对此次行动全力相助。就这样，锂锂一行人不仅获得了武器，还收获了几名实力强悍的帮手。

After Child Nickel explained everything, the soldiers decided to assist this mission. Thus, Lithium Li and his party not only obtained weapons, but also gained several powerful helpers.

等等！那不是……我们的老祖宗——初代镍镉电池吗？

Wait! Isn't that… our ancestor— the first nickel-cadmium battery?

没错，是我！不要紧张，放下枪。

That's right, it's me! Be relaxed and put down your gun.

我是通过时光穿梭而来。这次过来，是有一项意义重大的任务要交给你们……

I came here through a time machine. I'm here because I have an important mission for you…

我们的后辈——一个名叫镍霸的最新一代镍镉电池，在未来的时空里，做出了伤害同胞、危害电池王国发展的行为，作为先辈的我们有责任阻止他！

One of our descendants, a new-generation nickel-cadmium battery called Tyrant Nickel, hurts our people and puts the Battery Kingdom in danger. As his ancestors, it's our duty to stop him!

镍镉电池家族的年轻人们，请带上你们最厉害的武器，让他见识一下你们的强大！随我一起去劝他迷途知返吧！

Brave young nickel-cadmium batteries, grab your best weapons, and show him how powerful you are! Let's go together and guide him back to the right path!

好的！

Yes, Sir!

没想到的是，锂锂、闪闪、大铅刚从时空隧道中出来，就掉入了镍霸提前准备的陷阱之中。

To their surprise, Lithium Li, Shiny, and Big Lead fell into the trap set by Tyrant Nickel in advance as they came out from the space-time tunnel.

嘿嘿，没想到吧！看你们这次往哪儿逃！

Heh heh, didn't see that coming, did you? Let's see where you can run this time!

你……你怎么知道我们在这里？

How… How did you know we were here?

根据我的观察，你们每一次来到这个时空，都是在这个位置出现，所以我为你们精心准备了这个礼物！好好享受吧！

From my observation, every time you come to this time and space, you always appear at this location. That's why I prepared this gift for you! Enjoy it!

没等镍霸得意多久，只见时空隧道里竟飞出了两架直升机，将困在陷阱里的几人救了出去。原来，是镍镉电池家族的前辈们来支援了！

Before Tyrant Nickel could savor his triumph for long, two helicopters suddenly flew out of the tunnel, rescuing the heroes in the trap. It was the elders of the nickel-cadmium battery family coming to help!

赶紧交出宝石，向大家赔礼道歉！

Hand over the gem and say sorry to everyone right now!

你这个不肖子孙！我们镍镉电池家族的一世英名，就要葬送在你的手里了！

You're a disgrace to our family! You're about to ruin the great name of the nickel-cadmium battery family!

快！上来！

Quickly! Come up!

镍霸眼见不敌，竟开始发动手中的心灵宝石，企图用魔法控制住所有人。闪闪看出了他的意图，立即用空间宝石构建了一个空间保护罩，将众人与外界隔绝起来。

Realizing he was going to lose, Tyrant Nickel tried to use the Mind Gem to control everyone with magic. But Shiny saw his plan and quickly used the Space Gem to create a protective shield, keeping everyone safe inside and away from his magic.

老祖宗、前辈们，你们可误会我了！我所做的一切，都是为了复兴我们镍镉电池家族，重振家族当年的荣光啊！

My elderly ancestors, you've got it all wrong! Everything I've done is to bring back the glory of our nickel-cadmium battery family!

大家小心！

Be careful, everyone!

来，跟我一起见证这伟大的时刻吧！

Come and see this amazing moment with me!

发动宝石需要使用者消耗自身的电力，闪闪在力量宝石的加持下，拥有源源不断的电力来维持空间保护罩。

而手持心灵宝石的镍霸就无法长时间地消耗电力了。锂锂不忍心见到镍霸的电力被消耗殆尽，于是开始了劝降。

Activating the gems consumes the user's energy. Blessed by the power gem, Shiny has a steady stream of power to maintain the shield.

However, Nickel Tyrant, who held the Mind Gem, was unable to consume power for a long time. Lithium couldn't bear to see Nickel Tyrant's power being exhausted, so he began to persuade him to surrender.

镍霸，你是耗不过我们的，快点投降吧！归还宝石，回头是岸！

Tyrant Nickel, you can't keep this up forever. Just surrender! Return the gem and stop before it's too late!

投降？开什么玩笑？在这历史回廊里做个假国王也没什么意思……我要想办法回到电池王国……既然，锂锂这么想我投降，不如……我就借坡下驴！嘿嘿！

Surrender? Are you joking? Pretending to be a fake king in this Historical Corridor isn't fun… I need to find a way back to the Battery Kingdom. But since you're so eager to see me quit, Lithium Li… maybe I'll just play along! Heh heh!

就在锂锂苦口婆心地劝导时，镍霸突然停止了攻击，扑通一声跪倒在众人面前，开始忏悔。这180度的态度大转变令众人目瞪口呆。

As Lithium Li tried to persuade him, Tyrant Nickel suddenly stopped. He threw himself on his knees with a loud thud, and began to confess. This dramatic shift stunned everyone.

啊？！
What?!

我知道错了！对不起大家！
I know I was wrong! I'm so sorry, everybody!

这么轻易就知错了吗？
What? Really?

接着，镍霸向众人诚恳地道了歉，并交出了手中的心灵宝石。见此情景，众人便没有再计较，闪闪也拿出了电池项链，准备引导心灵宝石归位。

Then, he sincerely apologized and hands over the Mind Gem. Seeing this, the group forgave him, and Shiny took out the battery necklace, ready to restore the Mind Gem back to its rightful place.

其实，我只是想得到力量，复兴我们镍镉电池家族……

Honestly, I just wanted power to revive our nickel-cadmium battery family…

我也不想伤害你们……

I never meant to harm you…

没关系！谁都会犯错，你能及时改正就好！

It's alright! Everybody makes mistakes. What matters is that you've corrected it.

知错能改，就是好孩子！

If you can fix your mistakes, you're a good boy!

太好了！问题解决了！

Fantastic! It's fixed!

突然，镍霸猛地一跃而起，将心灵宝石和电池项链一并夺走！此时大家才明白，原来镍霸是在假意悔过！

紧接着，镍霸一边发动心灵宝石，命令被控制的电池人拦住众人，一边发动空间宝石，打开了时空隧道。

Then Tyrant Nickel jumped up suddenly and snatched away the Mind Gem and the battery necklace! It was only then that everyone realized that Tyrant Nickel had been pretending!

Following that, Tyrant Nickel activated the Mind Gem to command the controlled battery men to block everyone. At the same time, he activated the Space Gem and open the time tunnel.

哈哈——
Ha-ha—

哎呀！
Oops!

就在这千钧一发之际，镍子楚飞身扑向即将逃走的镍霸，用尽全身力气将他困在了原地。

In the nick of time, Child Nickel threw himself at the fleeing Tyrant Nickel and trapped him with all his strength.

哎呀！
Argh!

让开！你不会明白的！
Get out of the way! You just don't get it!

锂锂迅速反应过来，向镍霸发起了远程电力攻击，将其手中的电池项链精准地击落在地。

Lithium Li quickly launched a long-range electric attack on Tyrant Nickel, precisely knocking the battery necklace out of his hand to the ground.

看我的！
Leave it to me!

不巧的是，锂锂这一击，刚好击中了空间宝石，空间宝石也因此碎成两半。这使得原本打开的时空隧道变得极不稳定，入口开始快速缩小。

见此情形，镍霸顾不得别的，纵身跳入了即将关闭的时空隧道之中。在最后关头，还顺手捡走了其中一半空间宝石碎片。

Unfortunately, Lithium Li hit the Space Gem, breaking it in half. This made the time tunnel highly unstable, and its entrance began to shrink rapidly.

Tyrant Nickel jumped into the closing time tunnel without thinking. At the last moment, he grabbed one half of the Space Gem fragments.

一定要夺回宝石和项链！
We must get the gems and the necklace back!

你又欺骗了我们，你这个骗子！
You've tricked us again. You liar!

嘿嘿！咱们后会有期！
Hehe! Bye!

被心灵宝石所控制的电池人在脱离控制后，会有很长一段时间的后遗症。这不，镍霸一离开，原本被心灵宝石所操控的电池人就仿佛被抽走了灵魂一般，像木偶一样呆呆地立在原地，众人只得留在原地休养。

然而就在这时，锂锂收到了机器人X发来的一条紧急通知……

Freed from the control of the Mind Gem, the battery men would experience a long-term hangover. After Tyrant Nickel left, they seemed to have their souls taken away and stood there like puppets. Thus, everyone had to stay and rest.

However, at this moment, Lithium Li receives an urgent message from Robot X…

他们的安全阀都被破坏了，我先给他们简单维修一下吧。

Their safety vents have all been damaged. I'll perform some quick fixes.

不好了！电池王国出事了！锂锂，你快回来！

Oh no! The Battery Kingdom is in trouble! Lithium Li, you need to return right now!

电池大揭秘
Secrets behind Batteries

电的发现
The Discovery of Electricity

电，可以说是人类历史上最伟大的发现之一。有了电，人类就可以更有效地使用机械设备。

电的发现，使人类的科学文明有了一个质的飞跃。那么，人类究竟是怎么发现"电"的呢？

Electricity is considered one of the greatest discoveries in human history. With electricity, humans can operate machinery more effectively. The discovery of electricity marked a qualitative leap to scientific civilization. So, how did humans discover "electricity"?

古代中国人对"电"的认知
Ancient Chinese's Understanding of Electricity

事实上，中国人的祖先很早就发现了"电"，最早在距今有 3 000 多年历史的甲骨文中，就有对"电"的记载，它的"形状"和闪电相似。

In fact, Chinese ancestors discovered "electricity" very early. Records of electricity, resembling the shape of lightning, can be found in oracle bone script dating back more than 3 000 years.

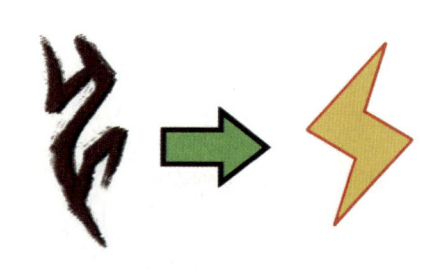

甲骨文的"电"字
Oracle Bone Script for "Electricity"

由东汉经学家、文字学家许慎编撰的《说文解字》记载："电，阴阳激燿也"，大意就是说：阴阳相激而产生的耀眼光芒，被称为"电"。

不过，那时候的古人说的"电"和我们如今所说的电，不是同一种电。我们说的电一般指的是电流，而古人说的"电"是闪电。

除了对闪电这一自然现象的观察记录以外，人类对电的认识其实是从静电开始的。古人常常将电与磁的现象归纳在一起，认为它们之间存在一种相互吸引的现象。

Shuowen Jiezi, an ancient Chinese dictionary compiled by philologist Xu Shen in the Eastern Han Dynasty, records that "Electricity stimulates yin and yang", which basically means the dazzling light produced by the excitation of yin and yang is called "electricity".

However, the term "electricity" in ancient times differs from how we understand it today. Modern "electricity" generally refers to electric currents, while the ancient term refers to lightning.

Apart from observations and records of lightning as a natural phenomenon, our understanding of electricity actually began with static electricity. Ancient people often grouped the phenomena of electricity and magnetism together, believing they were signs of mutual attraction.

静电，是一种处于静止状态的电荷或者说不流动的电荷（流动的电荷就形成了电流）。静电是通过摩擦或电荷的相互吸引使电荷重新分布而形成的。

Static electricity is a stationary charge that does not flow (the flowing charges can form currents). It is formed by the redistribution of charges through friction or the mutual attraction of charges.

西汉末年的《春秋纬·考异邮》中就有"瑇（玳）瑁吸芥"现象的记载。玳瑁是一种隶属于海龟科的爬行动物，在中国古代，其甲壳为制作饰品的珍贵材质；"芥"是指小的、轻微的物体。玳瑁甲壳制品平滑而有光泽，能够吸引微小物体，古人们发现了该现象，便将其记载下来。

In a text from the late Western Han Dynasty, *Spring and Autumn Weft: Examination of Anomalies*, there is a record of the phenomenon of "tortoiseshell (or hawksbill) attracting small objects". The hawksbill turtle, a type of sea turtle, had shells that were considered precious materials for making ornaments in ancient China. The term "jie" refers to small or light objects. The smooth and shiny hawksbill shell products can attract small objects, and the ancients documented this phenomenon upon discovering it.

王充的《论衡·乱龙篇》中，对此现象有进一步的记载："顿牟（即玳瑁）掇芥，磁石引针，皆以其真是，不假他类。他类肖似，不能掇取者，何也？气性异殊，不能相感动也。"

Wang Chong's *Lunheng: Chapter on the Chaotic Dragon* further records this phenomenon: "The hawksbill (i.e., tortoiseshell) attracts small objects and the magnet attracts needles due to their true nature, not relying on other types. Why can other objects, though similar to small seeds or steel needles, not be attracted? Because their 'qi' nature differs, preventing interaction."

这段话的意思是说，经过摩擦的玳瑁能吸引芥籽，磁石能吸引钢针，这是因为它们之间的"气性"相同，能相互感应而动；其他看起来与芥籽、钢针相似的东西，但因与玳瑁、磁石的"气性"不同，所以不能相互感应而动。

It means that a hawksbill shell that has been rubbed can attract small seeds, and a magnet can attract a steel needle because they share the same "qi" nature, allowing them to interact and move; other objects, though similar to small seeds or steel needles, cannot interact and move as they differ from the hawksbill or magnet in "qi" nature.

东晋的《山海经图赞》中，也有类似的记载，即"磁石吸铁，玳瑁取芥，气有潜通，数亦宜会。"也把静电和静磁并列。

在西晋时，张华撰写的《博物志》中也有关于静电的记载："今人梳头、脱著衣时，有随梳，解结有光者，亦有咤声。"意思是说在人们梳头、穿脱衣服时，常发生摩擦起电，有时还能看到小火花和听到微弱的响声。

In *Illustrations and Eulogies of Fantastic Creatures of the Mountains and Seas* from the Eastern Jin Dynasty, there is a similar record that "The magnet attracts iron, the hawksbill gathers small objects; their 'qi' has a subtle connection, and this connection aligns with the laws of nature". This also places static electricity and static magnetism side by side.

During the Western Jin Dynasty, Zhang Hua's *Book of Natural History* also contains records about static electricity: "Nowadays, when people comb their hair or take off their clothes, there are often shiny particles that follow the combing or the untying of knots, sometimes accompanied by faint sounds." This means that when people comb their hair or put on and take off clothes, static electricity often occurs with small sparks and sounds.

古希腊人对电的认知
Ancient Greeks' Understanding of Electricity

在距今约 2 600 年前的古希腊鼎盛时期，贵族妇女外出时都喜欢穿柔软的丝绸衣服，戴琥珀做的首饰。人们外出时，总把琥珀首饰擦拭得干干净净。

In the heyday of ancient Greece about 2 600 years ago, noble women liked to wear soft silk clothes and amber jewelry when they went out. They always wiped the amber jewelry clean.

但是，不管擦得多干净，它很快就会吸上一层灰尘。

虽然许多人都注意到了这个神奇的现象，但一时都无法解释它。古希腊哲学家泰勒斯，在经过仔细观察后，发现挂在脖子上的琥珀首饰在人走动时会不断晃动，频繁地摩擦身上的丝绸衣服，从而得到了启发——用毛皮摩擦琥珀后，琥珀就能吸引绒毛、麦壳屑、毛发等轻微的东西。

However, no matter how clean it was, it would soon absorb a layer of dust.

Although many people noticed this magical phenomenon, they could not explain it at that time. The ancient Greek philosopher Thales, after careful observation, found that the amber jewelry hanging on the neck would keep swaying when people walked, rubbing against the silk clothes, which inspired him. After rubbing amber with fur, amber can attract slight things such as fluff, wheat shell shavings, and hair.

因此，古希腊人认为在琥珀中存在一种特殊的神力，并把这种神力叫做"电"。

Therefore, the ancient Greeks believed that there was a special divine power in amber, which they called "electricity".

"照亮世界"的实验
Experiment of "Illuminating the World"

人类对电现象认识得很早，不过那时候大多是雷雨天气的闪电以及摩擦生电的衍生现象。千百年来，关于电的问题一直困扰着古人，他们将电分为"天电"和"地电"，我国古人还曾创作了"雷公电母"的神话，认为打雷、闪电都是天上的神发怒的表现。后来，越来越多的人开始研究电，直到1752年6月，出现了一个非常著名的实验——富兰克林风筝实验。

Humans have long been aware of electrical phenomena, though their knowledge was mostly limited to lightning during thunderstorms and the effects of frictional electricity. Our ancients have been puzzled by electricity for thousands of years. They divided it into "heavenly electricity" and "earthly electricity". Even a myth about the "Thunder God and Lightning Mother" was created by ancient Chinese, for they believed that thunder and lightning were signs of gods' anger. Later, more people began to study electricity. A notable experiment took place in June 1752, known as the Franklin Kite Experiment.

一位胆子巨大的美国科学家——本杰明·富兰克林，他用风筝去牵引空中的雷电，将电"捕捉"了下来，以此来证明那时候人们口中的"天电"与"地电"是一样的。

Benjamin Franklin, a most daring scientist in America, flew a kite to attract lightning from the sky, intending "to capture" the electricity to prove that "heavenly electricity" and "earthly electricity" were the same.

风筝实验的成功使富兰克林在全球科学界名声大振，那个时期，人们纷纷开始对富兰克林这个实验进行验证。

于是，在众多科学家的共同努力下，人类更进一步地知晓了电的性质，并延伸出了"电流"的概念。

The successful kite experiment earned Franklin worldwide recognition in the scientific community. Then, many people tried to verify his findings.

Thanks to the joint efforts of numerous scientists, humans further understood the nature of electricity and developed the concept of "currents".

电池王国起源之谜
The Mysterious Origin of the Battery Kingdom

在发现电之后，人类面临的问题就是如何利用电。

After the discovery of electricity, the challenge was how to use it.

世界上第一个电池
The World's First Battery

电池的诞生，是基于人们对于获取持续而稳定的电流的需要；电池的发明，是来源于一次青蛙的解剖实验所产生的灵感。

1780 年的一天，意大利解剖学家路易吉·加尔瓦尼在解剖青蛙时，两手分别拿着不同的金属器械，无意中金属器械同时碰到青蛙的大腿上，青蛙腿部的肌肉立刻抽搐了一下，仿佛受到电流的刺激，而如果只用一种金属器械去触动青蛙，就无此种反应。

The birth of the battery arose from the need for a consistent and stable current. It was inspired by a frog dissection experiment.

One day in 1780, when Luigi Galvani, an Italian anatomist, was doing a frog dissection, he held different metal instruments in both hands. Inadvertently, the metal instruments touched the frog's thigh at the same time, and the muscles of the frog's leg immediately twitched, as if stimulated by an electric current. If only one metal instrument was used to touch the frog, there would be no such reaction.

> 生物电！
> Bioelectricity！

> 电学的发展可离不开我们的启发！
> We played a key role in how electricity advanced!

加尔瓦尼认为，出现这种现象是因为动物躯体内部产生的一种电，他称之为"生物电"。加尔瓦尼的发现引起了物理学家们的极大兴趣，他们重复着伽伐尼的实验，试图找到一种产生电流的方法。

意大利物理学家亚历山德罗·伏特也受到了启发，他在多次重复加尔瓦尼的实验之后认为：青蛙的肌肉之所以能产生电流，是因为肌肉中某种液体在起作用。

为了验证自己的观点，伏特把一块锌板和一块锡板浸在盐水里，发现连接两块金属的导线中有电流通过。

Galvani believed this phenomenon happened because of a kind of electricity generated inside the animal, which he called "bioelectricity". Galvani's findings sparked the interest of physicists. They repeated his experiment and tried to find a way to generate currents.

Italian physicist Alessandro Volta was also inspired by Galvani's work. After conducting similar experiments many times, he thought that the currents in the frog's muscles resulted from some liquid inside.

To test his hypothesis, Volta put a zinc plate and a tin plate in salt water and discovered that currents flowed through the wire connecting the two metals.

于是，他就把 30 片圆锌片和 30 片圆铜片相互叠成一堆，在每片之间又夹入一片浸有浓盐水的吸水纸，然后，他从铜片和锌片上各引出一根导线，将两根导线相接，竟有放电的火花产生。

So, he stacked 30 round zinc sheets and 30 round copper sheets, with a piece of absorbent paper soaked in saltwater between each pair. Then, he took a wire from each of the copper and zinc pieces, and connected the two wires, and there was a spark of discharge.

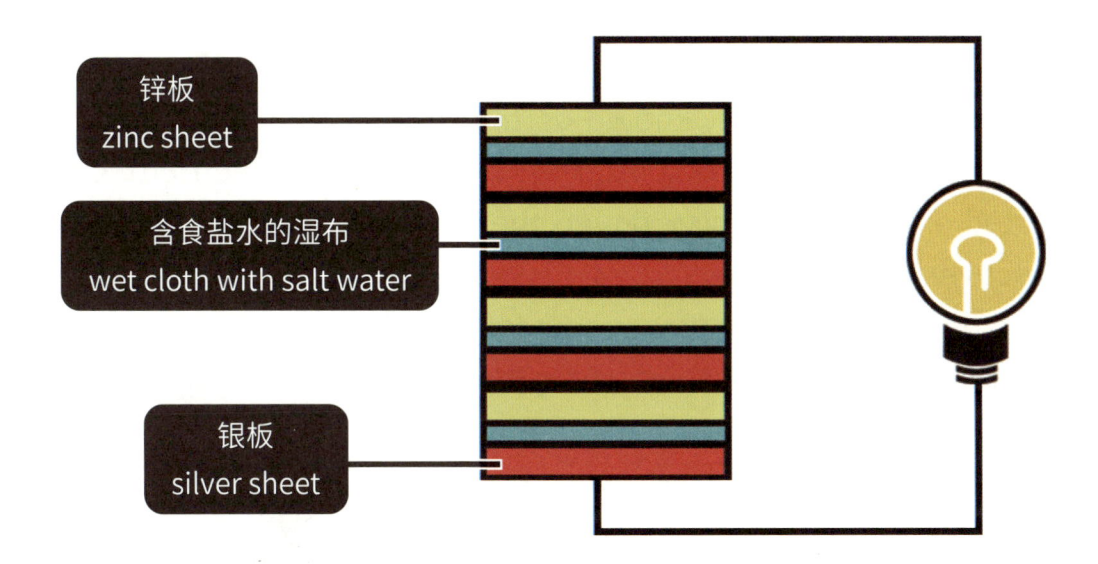

锌板
zinc sheet

含食盐水的湿布
wet cloth with salt water

银板
silver sheet

电池小知识
Battery tip

伏特是国际单位制中表示电压的基本单位，简称"伏"，符号 V。该单位是为了纪念意大利物理学家亚历山德罗·伏特而命名的。

Volt (V) is the basic unit of voltage in the International System of Units. It was named after the Italian physicist Alessandro Volta.

最终，伏特得出了结论：两种金属片中，只要有一种与溶液发生了化学反应，金属片之间就能够产生电流。

基于这个理论，在1799年，伏特成功制成了世界上第一个电池——伏打电堆。这个伏打电堆实际上就是串联的电池组，它的出现证明了电是可以被人为制造出来的，使许多新的实验和发现成为可能。

化学电源就此出现，这为现代电池技术打下了基础，开创了电化学发展的新时代。

Volta concluded that currents could be produced between two metal sheets as long as one of them reacted chemically with the solution.

In 1799, based on this principle, Volta successfully created the first battery in the world, known as the Voltaic Pile. This device was simply a series of connected batteries. It proved that electricity could be generated artificially. It also paved the way for numerous experiments and discoveries.

The birth of chemical power sources laid the groundwork for modern battery technology and opened a new era in electrochemistry.

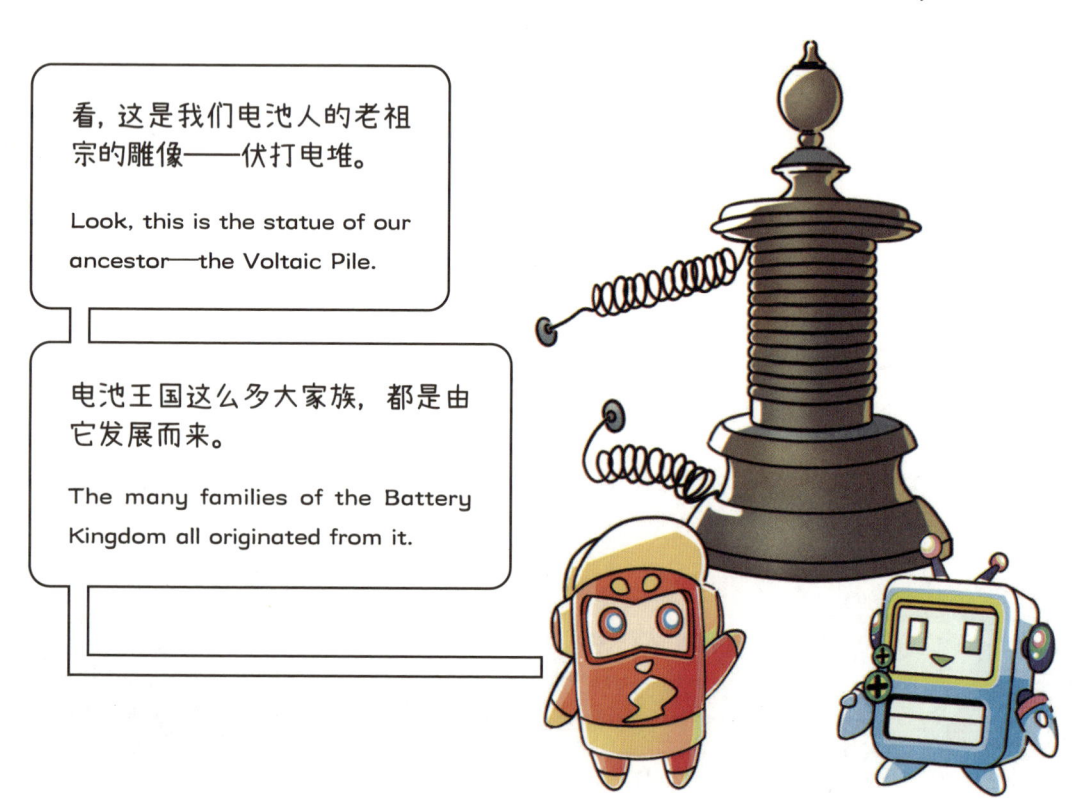

看，这是我们电池人的老祖宗的雕像——伏打电堆。

Look, this is the statue of our ancestor—the Voltaic Pile.

电池王国这么多大家族，都是由它发展而来。

The many families of the Battery Kingdom all originated from it.

世界上第一台电动机
The First Electric Motor in the World

有了伏打电堆，人们开始研究电流产生的各种效应，并对"电有什么作用"展开了广泛研究。

1821年，英国科学家迈克尔·法拉第发明了世界上第一台电动机。这项轰动世界的重大发明，初步证实了使用电流可以让物体运动起来。

With the birth of the Voltaic Pile, researchers began to explore the effects of currents and conducted extensive research on the question, "What can electricity do?".

In 1821, British scientist Michael Faraday created the world's first electric motor. This great invention shocked the world and provided initial proof that electric current could make objects move.

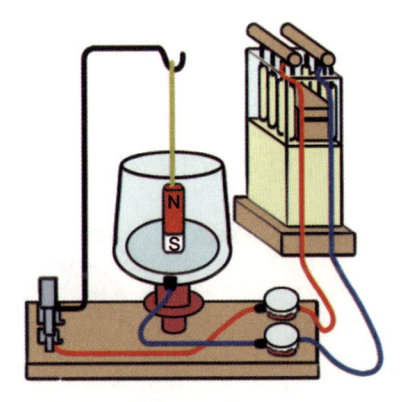

此后，各种各样的电器开始逐渐浮现于世，人类由此进入"电气时代"，电池也开启了不断进化的历程，衍生出了各个大家族，构成了如今强大的电池王国。

Following this, various electrical devices began to emerge, marking the dawn of the "Electric Age". Batteries also started to evolve and derived various large families, forming the robust Battery Kingdom we know today.

话说干电池
The Story of Dry Battery

整个电池王国的发展史也可以说是一个"尝试用各种金属造电池"的历史。实际上，只要有两种金属浸泡在某种溶液中，就有可能产生电池效应。例如，接受过金属补牙手术的人们会发现，用舌头去舔补牙的金属，会有"麻麻"的感觉，就是因为补牙用的多种金属在口腔中产生了电池效应。

The history of the Battery Kingdom can also be viewed as a story of "trying to make batteries with various metals".

In fact, a battery effect can occur as long as two different metals are submerged in a solution. For example, people with metal dental fillings may feel a "tingling" when they lick their fillings. This happens because the different metals in the filling create a battery effect inside the mouth.

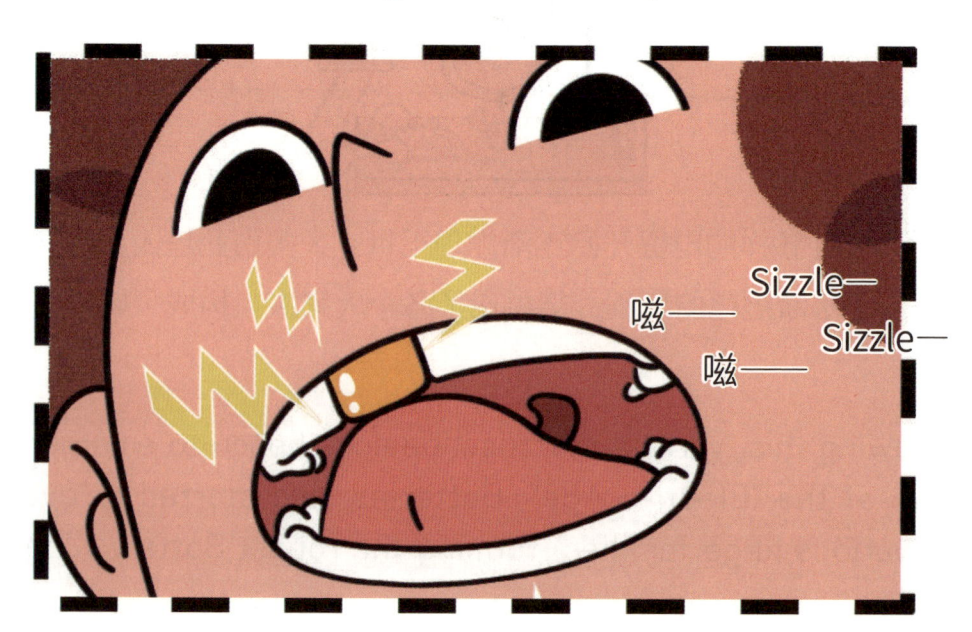

自伏打电堆问世以来，科学家们在这个基础上进行实验，又陆续研发出了更多效果更好的电池。

然而在当时，无论哪种电池，都需要在两个金属板之间灌装液体。这样的电池搬运起来很不方便，特别是蓄电池，所用的液体是硫酸，在挪动时很危险。

Since the birth of the Voltaic Pile, scientists have carried out experiments on this basis and developed more and better batteries.

However, at that time, no matter what kind of battery it was, a tank of liquid was required between two metal plates. It was very inconvenient to transport such batteries, especially the dangerous storage batteries which used sulfuric acid as the liquid.

看，那时候科学家们对电池的研究仍在探索阶段，电池的可移动性非常差。

Look, back then, scientists were still in the exploratory stage of battery research, and the mobility of batteries was very poor.

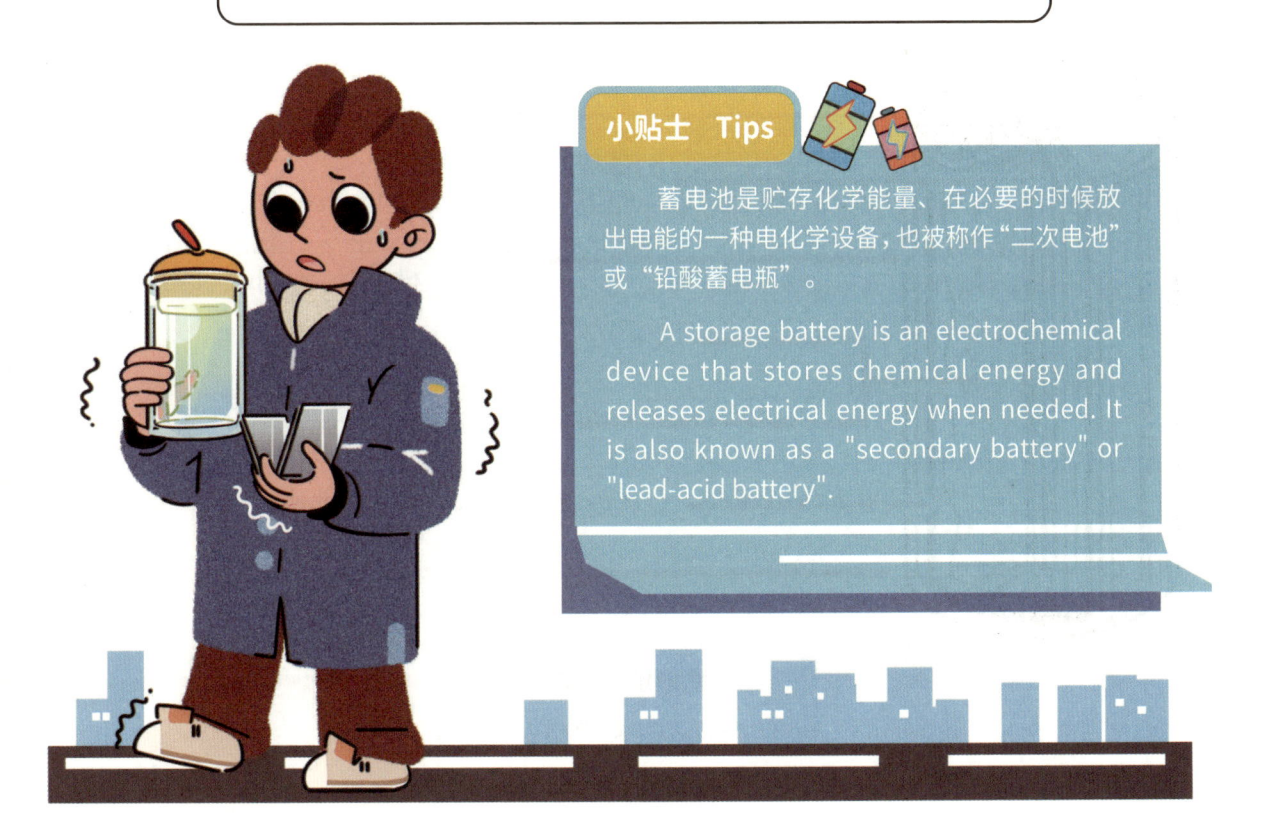

小贴士 Tips

蓄电池是贮存化学能量、在必要的时候放出电能的一种电化学设备，也被称作"二次电池"或"铅酸蓄电瓶"。

A storage battery is an electrochemical device that stores chemical energy and releases electrical energy when needed. It is also known as a "secondary battery" or "lead-acid battery".

终于，在 1860 年，"干"性电池出现了。勒克朗谢制出了原电池的雏形，发明了碳锌干电池，这种电池是用糊状电解质取代了此前潮湿、水性的电解液。相对于液体电池而言，干电池的电解液为糊状，不会溢漏，便于携带，因此获得了广泛应用。它可以说是如今广受人们欢迎的碱锰电池的"祖先"了。

真正意义上的干电池出现于 1886 年前后，赫勒森在碳锌干电池的基础上进行改进，并最终发明了我们今天用的干电池。

Finally, in 1860, the "dry" battery appeared. Leclanche made the prototype of the primary battery, inventing the carbon-zinc dry battery, which replaced the previously moist, water-based electrolyte with a paste electrolyte. Compared to liquid battery, dry battery features a paste electrolyte that doesn't leak and is easy to transport, leading to their wide use. They are the "ancestors" of today's popular alkaline-manganese batteries.

The real dry battery was developed in around 1886 when Hellesen improved the carbon-zinc dry battery and created the dry battery we use today.

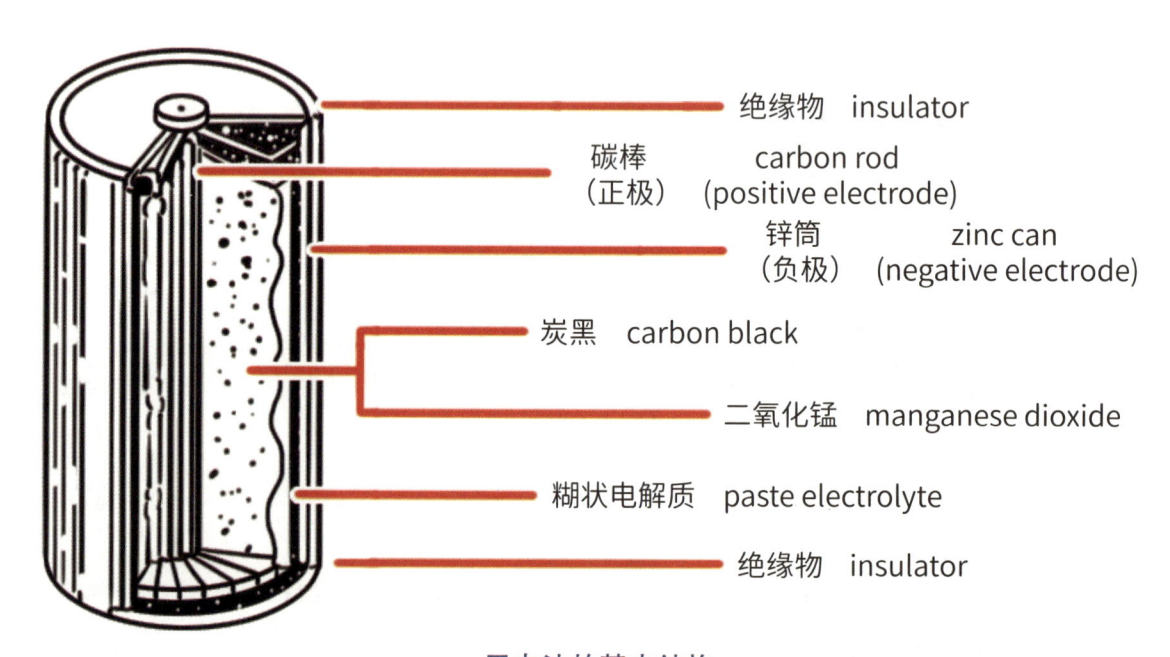

绝缘物 insulator
碳棒（正极） carbon rod (positive electrode)
锌筒（负极） zinc can (negative electrode)
炭黑 carbon black
二氧化锰 manganese dioxide
糊状电解质 paste electrolyte
绝缘物 insulator

干电池的基本结构
basic structure of a dry battery

随着科学技术的进步，干电池已经发展成为一个庞大的家族，现如今已经有100多种。常见的有普通锌-锰干电池、碱性锌-锰干电池、镁-锰干电池、锌-空气电池、锌-氧化汞电池、锌-氧化银电池、锂-锰电池等。

不过，最早发明的碳锌电池依然是现代干电池中产量非常大的电池。

With advancement in science and technology, batteries have grown into a large family with over 100 types today. Common varieties include standard zinc-manganese dry batteries, alkaline zinc-manganese dry batteries, magnesium-manganese dry batteries, zinc-air batteries, zinc-mercury oxide batteries, zinc-silver oxide batteries, and lithium-manganese batteries, etc…

However, the earliest carbon-zinc battery is still one of the most widely produced batteries in modern dry batteries.

> 在我被做成一次电池时，我们同属于一次电池家族中的干电池分支，又都是"锌"系电池，因此你可是我的老前辈！
>
> When I was created as a primary battery, we belonged to the dry battery branch as well as the 'zinc' battery. You are my senior, after all!

> 哈哈哈，锌博士，你怎么变成这样了？
>
> Ha-ha-ha, Dr. Zinc, how did you become like this?

话说蓄电池
The Story of Storage Batteries

19 世纪末，许多电器都已经诞生，如电灯、电话、电报、电唱机等。这些电器的问世，为人们的生活带来了便利和乐趣。

At the end of the 19th century, many electrical appliances have been born, such as light bulbs, telephones, telegraphs, record players, etc. The advent of these electrical appliances has brought convenience and fun to people's lives.

电灯泡
light Bulb

电报机
telegraph

专线电话机
private-line telephone

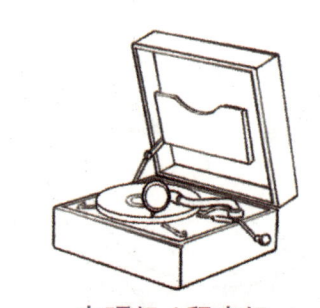

电唱机 / 留声机
record player /gramophone

电器都是靠电力驱动工作的，没有了电，这些东西就毫无价值。

在那时，电的来源有两个途径：一是由发电机发电，二是由蓄电池供电。

蓄电池有一个独特之处——当电池使用一段时间、电压下降后，可以给它通以反向电流，使电池电压回升。因为这种电池能够充电，并且能被反复使用，所以被称为"蓄电池"。

蓄电池比发电机更加小巧，便于携带。有了它，偏僻的地方也可以用上电。

All these gadgets are powered by electricity. Without electricity, they are worthless.

Back then, electricity came from two main sources: generators and storage batteries.

Storage battery has a unique feature—when the battery is used for a period of time and the voltage drops, it can be given a reverse current to make the battery voltage rise. Because this battery can be charged and used repeatedly, it is called "storage battery".

Compared to generators, storage batteries are much smaller and portable. With them, you can have electricity even in remote places.

世界上第一个铅酸蓄电池
The First Lead-acid Battery in the World

1859 年，普朗特发明出用铅和硫酸做的电池。这种电池就是铅酸蓄电池的前身，它便于携带，使用方便，也是最早的二次电池。

In 1859, Plante invented a battery made of lead and sulfuric acid. This battery is the predecessor of lead-acid batteries, which is convenient to carry and use, and is also the earliest secondary(rechargeable) battery.

前辈好！
Hello, senior!

看到你现在的样子，我感到很欣慰。
I'm relieved to see you in this state.

不过，虽然它解决了电池只能一次性使用的问题，但新的麻烦也随之而来——它的使用寿命太短，甚至只有一个多小时，人们称它为"短命蓄电池"。这导致蓄电池虽然在技术上实现了巨大突破，但实用性仍然不高。

However, while it solved the problem of one-time-use batteries, there came a new issue: its lifespan was very short—even just over an hour. People called it the "short-lived battery". Though it marked a huge technological breakthrough, it wasn't very practical yet.

我体内的硫酸和铅用完了，我也就"没命"了。
Once I run out of sulfuric acid and lead in my body, I'm "dead".

硫酸
sulfuric acid

铅
lead

电池小知识
Battery tip

普朗泰发明的电池，是用铅作为电极材料、浓硫酸作为电解质溶液制成的。它的工作原理是让铅和硫酸接触，发生电化学反应。在反应过程中，极化作用和电子转移就产生了电流。

但由于浓硫酸的腐蚀性极强，不久铅就被严重腐蚀，也就不能产生电流了。

The battery Plante invented is made of lead as an electrode material and concentrated sulfuric acid as an electrolyte solution. It works by bringing the lead into contact with sulfuric acid and an electrochemical reaction occurs. During the reaction, polarization and electron transfer generate an electric current.

However, due to the strong corrosiveness of concentrated sulfuric acid, the lead will soon be severely corroded and no current can be generated.

爱迪生与蓄电池
Thomas Edison and Storage Battery

托马斯·爱迪生，这位在当时已经发明了不少电器的美国科学家，意识到了解决蓄电池"短命"问题的重要性——如果不延长蓄电池的供电时间，将会影响蓄电池的推广使用。

Thomas Edison, an American scientist who had already invented many electrical appliances at the time, realized the importance of solving the problem of the "short life" of the battery. If the power supply time of the battery were not extended, it would affect the widespread use of the battery.

功夫不负有心人，爱迪生在 1890 年终于制成了可充电的镍铁蓄电池。经过长达 20 年、历经数万次的改进之后，终于在 1910 年实现了可充电镍铁蓄电池的商业化生产。为了纪念爱迪生的辛勤劳动，人们将镍铁蓄电池称为"爱迪生蓄电池"。

Hard work paid off. In 1890, Edison finally made a rechargeable nickel-iron storage battery. After 20 years and tens of thousands of improvements, in 1910, the battery was finally ready for commercial production. To honor his hard work, people started calling it "the Edison battery".

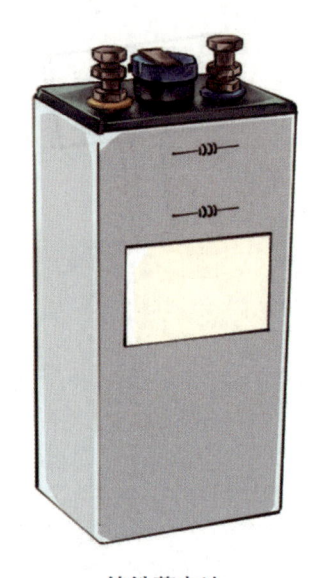

镍铁蓄电池

nickel-iron storage battery

镍铁蓄电池和当时的铅蓄电池相比，具有更高的能量密度，并且充电时间可缩减一半左右，一经发明就被认为是极有竞争力的化学电源之一，在重工业领域曾经风靡一时。

20 世纪 90 年代后期，随着电池王国的蓬勃发展，各个电池家族间的竞争也愈演愈烈，由于镍铁电池自身某些性能还不理想，它几乎被人们遗忘。近年来，由于它廉价、环保、安全等优点，镍铁蓄电池在许多领域的应用仍有相当大的发展空间，继续探索开发以改良它的性能，成为科研人员们目前紧要的工作之一。

Compared to the lead-acid battery of the time, the nickel-iron storage battery had a higher energy density and its charging time was reduced by about half. Once invented, it was considered one of the most competitive chemical power sources and became popular in heavy industry.

In the late 1990s, with the vigorous development of the battery kingdom, the competition among various battery families also intensified. Due to the unsatisfactory performance of some nickel-iron batteries, they were almost forgotten. There are still considerable room for them to be developed in many fields because of their features of being cheap, environmentally friendly and safe. Continuing to explore ways to improve its performance has become one of the urgent tasks of researchers.

爱迪生与电动汽车
Thomas Edison and the Electric Car

人们都说 21 世纪最伟大的发明是新能源汽车。但事实上，随着蓄电池技术的发展，电动汽车也随之出现，它的发展史甚至比燃油汽车的历史还要早半个世纪开始。

It is believed that the new energy vehicle is the greatest invention of the 21st century. In fact, the electric vehicle is also a great invention with the development of storage battery technology, which came into being even half a century earlier than that of the fuel vehicle.

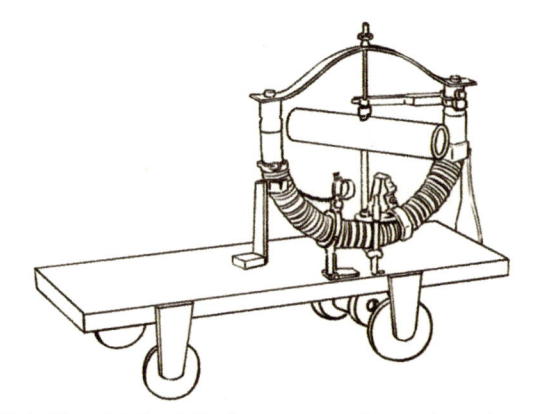

世界上第一辆电动汽车　　the first electric car

早在 100 多年前，发明大王爱迪生就用自己发明的镍铁蓄电池制造出了电动汽车，时速 20 英里，约 32 千米/时。1910 年，人们驾驶着爱迪生发明的电动汽车，从美国纽约行驶到了新罕布什尔，据说整个行驶距离达到 170 英里，约 273 千米。

Over 100 years ago, Edison made the world's first electric car with nickel-iron battery he had invented. It could reach a speed of 20 miles per hour (about 32 km/h). In 1910, people drove the car from New York to New Hampshire in the United States. It was said that the total distance traveled reached 170 miles, or about 273 kilometers.

"Edison brand" new battery can make the car go 1 000 miles!

蓄电池的发展现状
The Current State of Storage Battery

如今，蓄电池的种类越来越丰富，形式也越来越多样，从最早的铅蓄电池、铅晶蓄电池，到铁镍蓄电池以及银锌蓄电池，如今发展到铅酸蓄电池、太阳能电池以及锂电池等。

与此同时，蓄电池的应用领域越来越广，电容越来越大，性能越来越稳定，充电也越来越便捷。

Nowadays, the types of rechargeable batteries are becoming more and more diverse, from the earliest lead batteries, lead-crystal batteries, to iron-nickel batteries and silver-zinc batteries, and now to lead-acid batteries, solar cells, and lithium batteries.

At the same time, storage batteries are used in more and more places. They can store more power, work more steadily, and are easier and faster to charge.

话说太阳能电池
The Story of Solar Battery

太阳的光辉普照大地，它是光明的使者。地球上几乎所有生物利用的能量都直接或者间接来自太阳。

As the messenger of light, the sun shines brightly on Earth. Nearly all life on our planet gets its energy directly or indirectly from the sun.

太阳能电池的起源
The Origin of Solar Battery

人类享受着太阳带来的热能，但是在早期，人们还没发现太阳光能转化为电能。人类真正将太阳的热辐射作为一种能源加以利用，要从1615年开始算起。在那一年，法国工程师所罗门·德·考克斯发明了世界上第一台由太阳能驱动的发动机。

Humans enjoy the heat from the sun. But in the early days, it was not discovered that sunlight could be converted into electricity. Humans have really used the sun's thermal radiation as an energy source since 1615. That year, French engineer Solomon de Cox invented the world's first solar-powered engine.

随后，世界上又陆续出现了更多的太阳能动力装置和其他太阳能装置，如太阳能发动机、水泵等。

在这一阶段，人们对太阳能的研究重点是将太阳能直接转化为机械能，虽然比较实用，但当时的技术不够成熟，能量转化成本高居不下。

Following that, solar-powered devices and other solar energy devices such as solar engines and solar water pumps were developed.

During that period, researches focused mainly on converting solar energy into mechanical power. Although practical, the technology was still in its early stages, and the cost of conversion was high.

太阳能电池的发展
The Development of Solar Battery

直到1931年，第二次世界大战（以下简称"二战"）爆发，战争背景下人们对能源的需求更加迫切，当时的太阳能技术还不能解决能源短缺的问题。相较于将太阳能转化为动力来源，使用煤炭、石油和天然气等矿物燃料的性价比更高。因此，那个时候的太阳能研究工作逐渐受到冷落，进入了低潮期，参加研究工作的人数和研究项目都大量减少。

It was not until 1931, when World War II (hereinafter referred to as "WWII") broke out, that people's demand for energy became more urgent, and solar technology at that time could not solve the problem of urgent energy needs. Compared with converting solar energy into power sources, the use of fossil fuels such as coal, oil and natural gas is more cost-effective. Therefore, solar energy research at that time was gradually neglected and entered a low ebb with a significant decrease in the number of people participating in research and research projects.

在二战结束后的 20 年中，一些有远见的人士已经注意到石油和天然气等自然资源在迅速减少，科学家们开始呼吁人们重视这一问题，从而逐渐推动了太阳能研究工作的恢复和开展，太阳能研究热潮再次兴起。

在此期间，匈牙利发明家玛丽亚·泰尔凯什参与了多个太阳能项目，对太阳能的研究与发展作出了卓越贡献，因而被冠以"太阳能女王"的称号。

In the following 20 years after the WWII, some visionaries began to notice the rapid depletion of natural resources like oil and natural gas. Scientists started calling attention to this issue, which gradually led to a revival of solar energy research and a renewed surge in solar energy development.

During that period, Maria Telkes, a Hungarian inventor, participated in multiple solar energy projects and made outstanding contributions to the research and development of solar energy, earning her the title of "Solar Queen".

阳光迟早会被用作能源，那我们还等什么？

Sunlight will eventually be used as energy—what are we waiting for?

"太阳能女王"——发明家玛利亚·泰尔凯什
"Solar Queen" Maria Telkes, the inventor

太阳能？电能？　Solar Energy? Electricity?

太阳光转化为电能的标志性事件是"光伏效应"的发现。

1839年，法国科学家贝安托万·贝克勒尔发现光照射到硅材料上会引起"光起电力"现象。这种现象后来被称为"光生伏打效应"，简称"光伏效应"。

The discovery of the "photovoltaic effect" was a landmark event in the conversion of sunlight into electricity.

In 1839, French scientist Atoine Becquerel observed that light shining on a silicon material caused "photoelectricity" phenomenon. This phenomenon was later called the "photovoltaic effect".

电池小知识 Battery tip

光伏效应指光照使不均匀半导体或半导体与金属结合的不同部位之间产生电位差的现象。

光伏效应首先是由光子（光波）转化为电子、光能量转化为电能量的过程；其次是形成电压的过程。有了电压，就像筑高了大坝，如果电能量和电压之间连通，就会形成电流的回路。

The photovoltaic effect refers to the phenomenon in which light causes a potential difference between an uneven semiconductor or between different parts of a semiconductor and a metal.

It is a process of converting photons (light waves) into electrons and light energy into electrical energy, followed by the formation of voltage. Having voltage is like building a dam. If the electrical energy and voltage are connected, a current loop will be formed.

基于这个理论，随着科学家们对半导体物理性质的逐渐了解，以及加工技术的进步，美国的贝尔实验室在 1954 年研制出了第一个有实用价值的硅太阳能电池。

Building on this principle, as scientists gradually understood the physical properties of semiconductors and the manufacturing technology was improved, Bell Labs in the United States produced the first practical silicon solar battery in 1954.

太阳能电池
solar battery

一直到今天，太阳能电池的基本结构和工作原理都没有发生改变。

Even today, the basic structure and working principle of solar battery have not changed.

为什么要发展太阳能电池?
Why Develop Solar Battery?

碳捕获与封存技术
carbon capture and storage technology
5%

生物质能
biomass energy
7%

水电
hydropower
6%

核能
nuclear energy
3%

太阳能光伏发电
solar photovoltaic power
20%

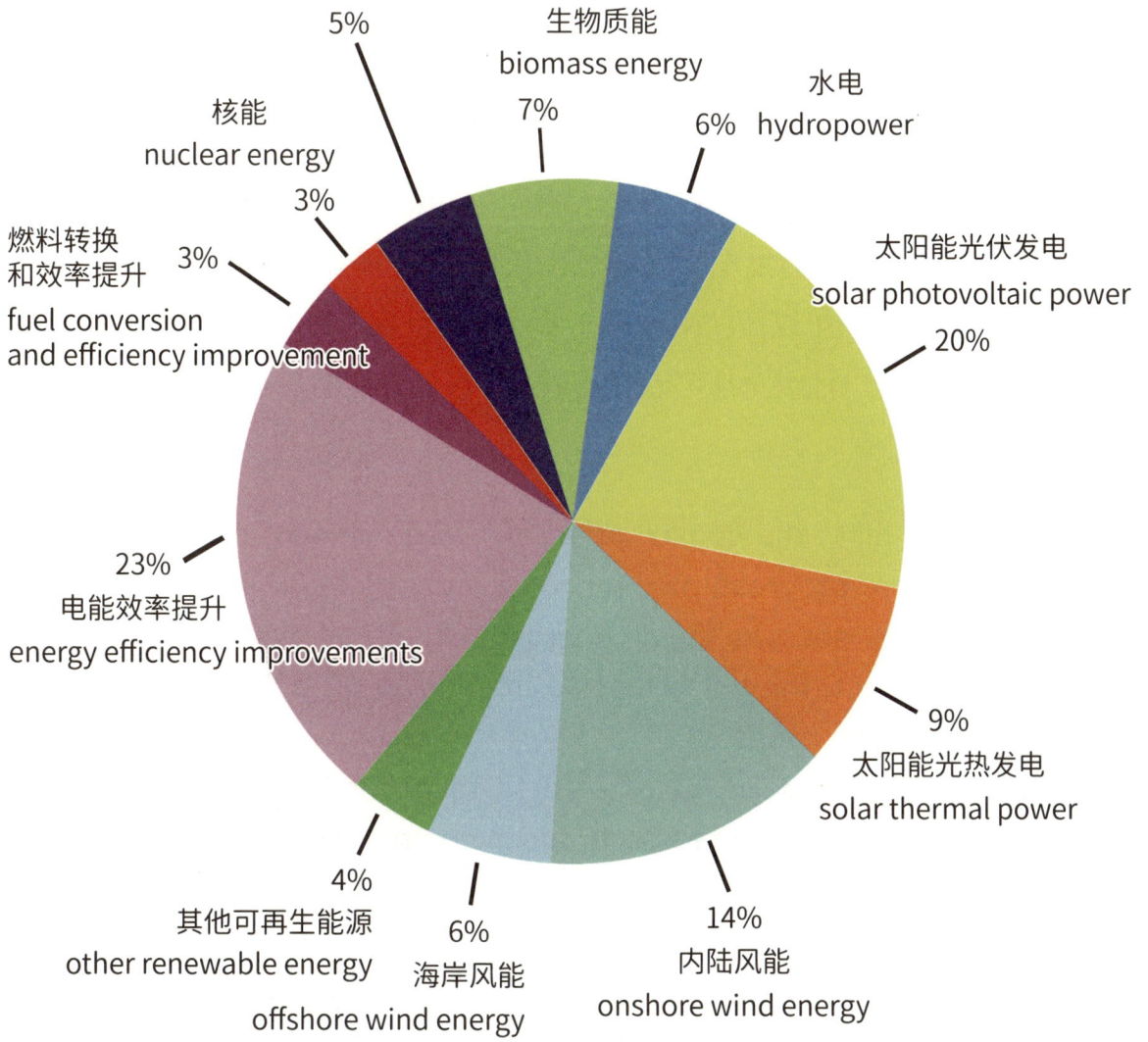

燃料转换和效率提升
fuel conversion and efficiency improvement
3%

23%
电能效率提升
energy efficiency improvements

9%
太阳能光热发电
solar thermal power

4%
其他可再生能源
other renewable energy

6%
海岸风能
offshore wind energy

14%
内陆风能
onshore wind energy

2014—2050 年不同能源技术对碳排放减排贡献的预测情况对比
(引自《新能源材料科学与应用技术》,《新能源材料科学与应用技术》编委会,科学出版社)

Comparison of the predicted contributions of different energy technologies to carbon emission reduction from 2014 to 2050
(quoted from *New Energy Materials Science and Application Technology*, *New Energy Materials Science and Application Technology*, editorial board, Science Press)

长期以来，大量化石能源的消耗推动着人类工业时代的进步。但是，传统能源（化石能源，如石油、煤炭等）在促进社会发展的同时，也导致空气污染日益加剧，全球温室气体排放量持续攀升。

人们为改变这个现状，作出了很多努力，目前的工业技术越来越注重对节能减排的贡献。

太阳能电池是太阳能光伏发电最核心的器件，伴随着光伏市场的逐步扩大、光伏技术的不断提高、光伏发电成本的日益下降和社会对清洁能源的迫切需要，它作为新型能源的优势越发明显。

太阳能电池具有永久性、清洁性和灵活性三大优点。

首先，太阳能电池寿命长。只要太阳存在，太阳能电池就可以一次投资而长期使用。其次，与火力发电、核能发电相比，太阳能发电不会引起环境污染。最后，太阳能电池还可以大、中、小并举，大到百万千瓦的中型电站，小到只供一户用的太阳能电池组，都可以灵活布置，这是其他电源无法比拟的。

For a long time, the consumption of a large amount of fossil energy has promoted the progress of the human industrial age. However, while traditional energy sources (fossil energy, such as oil, coal, etc.) promote social development, they also lead to increasing air pollution and global greenhouse gas emissions.

People have made a lot of efforts to change this situation, and industrial technology is increasingly focusing on energy conservation and emission reduction.

Solar battery is the core component of solar photovoltaic power generation. Its advantages as a new type of energy are becoming increasingly apparent with the gradual expansion of the photovoltaic market, the continuous improvement of photovoltaic technology, the decreasing cost of photovoltaic power generation, and the urgent need for clean energy in society.

Solar battery has three major advantages: permanence, cleanliness, and flexibility.

Firstly, solar battery has a long lifespan. As long as the sun exists, solar battery can be invested once and used for a long time. Secondly, compared to thermal and nuclear power generation, solar battery will not cause environmental pollution. Finally, solar battery can come in different sizes. They range from medium-sized power plants that produce millions of kilowatts to small groups of battery that only power one household. This makes solar power different from other energy sources.

我国太阳能电池的发展现状
Current Development of Solar Battery in China

现在，我国的光伏产业已经达到世界领先水平，为经济社会发展和生态环境保护共赢局面的促成保驾护航。

Now, our country's photovoltaic industry has reached the world's leading level, escorting the win-win situation of economic and social development and ecological and environmental protection.

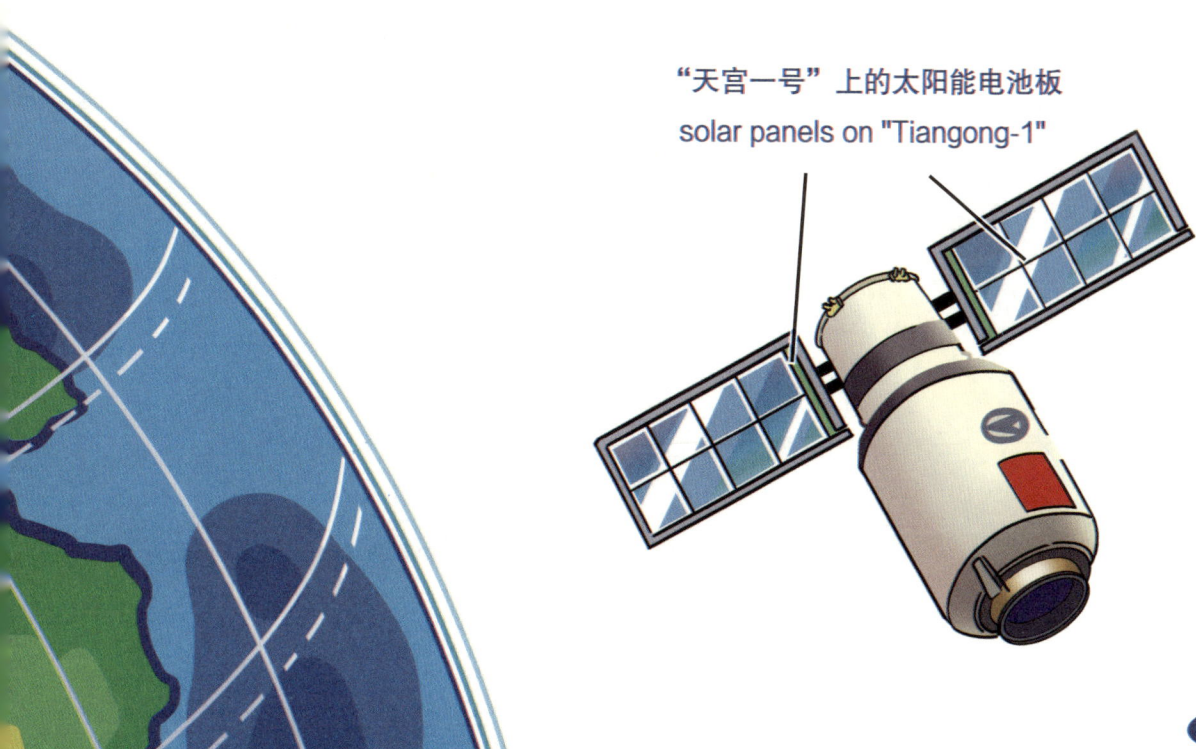

"天宫一号"上的太阳能电池板
solar panels on "Tiangong-1"

中国作为新的世界经济发动机，光伏产业呈现出前所未有的活力。

我国的太阳能电池经历了从无到有、从空间站到地面、由军到民、由小到大、由单品种到多品种，以及光电转换效率由低到高的艰难而辉煌的历程，大量的光伏企业应运而生。

As a new engine of the world economy, China's photovoltaic industry has shown unprecedented vitality.

China's solar battery has gone through a difficult but brilliant journey. They started from scratch. Being applied from space stations to the ground, they shifted from military use to civilian use. They grew from small-scale applications to large-scale ones, from a single type to multiple types, and their efficiency in converting sunlight to electricity has greatly improved. As a result, many photovoltaic companies have emerged.

话说锂电池
The Story of Lithium Battery

我们日常说的锂电池通常指的是锂离子电池，它和我们的生活联系非常密切，手机、电脑、电动汽车都离不开它。

现在的锂电池给我们的印象是体积不大、充电快、能量足，然而在早期，它并不是这个样子。

When it comes to lithium battery, it refers to lithium-ion battery (Li-ion), which is essential in our daily lives, powering everything from mobile phones to laptops and electric vehicles.

The current lithium battery strikes us as small in size, fast to charge, and rich in energy. In the early days, however, it was not what it is now.

Advanced Representatives of the Secondary Battery Family

"锂"想热潮
The Lithium Craze

锂离子电池由锂金属电池发展而来，它从诞生到被广泛应用，经过了许多科研工作者的不懈努力。回顾整个过程，可以说是非常曲折。

早在1817年，"锂"这种金属元素就被发现了，并且人们很快意识到锂金属的理化性质使其非常适合作为电池材料。

Lithium-ion battery was developed from lithium metal battery. It has undergone the unremitting efforts of many scientific researchers ever since its birth. Looking back at the whole process, it can be said to be very tortuous.

As early as 1817, the metal element "lithium" was discovered, and people quickly realized that the physical and chemical properties of lithium metal were very suitable for use as battery materials.

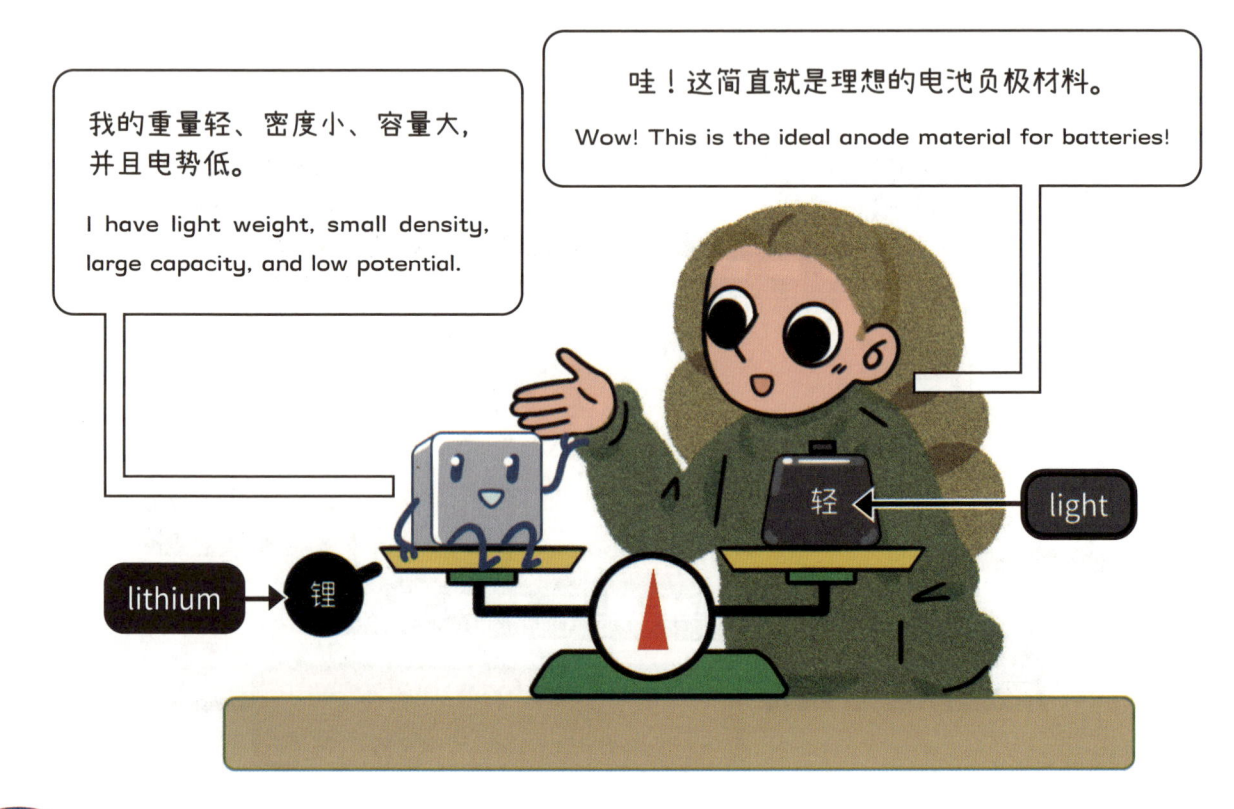

我的重量轻、密度小、容量大，并且电势低。
I have light weight, small density, large capacity, and low potential.

哇！这简直就是理想的电池负极材料。
Wow! This is the ideal anode material for batteries!

lithium → 锂

轻 ← light

锂是目前已知的世界上最轻的金属，也是密度最小的金属，它能浮于水面上。锂是电位最负的金属，是目前已知元素中金属活动性最强的，也是电化当量最大的金属。

因此，锂是极好的电池材料，由锂制成的电池的比能量最高。

不过，金属锂太活泼了，太容易和其他物质发生反应。不论是在水里，还是在煤油里，它都会浮上来与水或空气中的氧气发生化学反应。

Lithium is currently the lightest and least dense metal known in the world, and it can float on water. It is the metal with the most negative potential, the most active metal among all known elements, and the metal with the highest electrochemical equivalent.

Therefore, lithium is an excellent battery material, and batteries made of lithium have the highest specific energy.

However, metal lithium is so active that it can react with other substances very easily. Whether in water or kerosene, it will come up and react chemically with oxygen in water or air.

快来救救我！
Help! Help!

❌ 当锂遇到水……
when lithium meets water…

❌ 当锂遇到煤油……
when lithium meets kerosene…

对于这样一个顽皮的家伙，如何安全储存它是个大问题。科学家们经过研究，最后只好把它强行按进凡士林油或液体石蜡中，以隔绝空气储存。

For such a mischievous guy, how to safely store it is a big issue. After research, scientists decided to forcibly insert it into Vaseline oil or liquid paraffin to isolate it from the air.

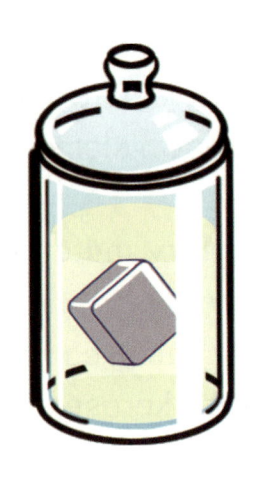

✓ 当锂遇到凡士林油或液体石蜡……
when lithium meets Vaseline oil or liquid paraffin…

锂的保存、使用或加工都比其他金属要复杂得多，导致这种金属在被发现后相当长的时间里都没有得到应用。在那个时期，锂的命运似乎被永远地封印在了实验室里。

不过，总有科学家不甘心。1913 年，美国的两位化学物理科学家——吉尔伯特·路易斯和弗雷德里克·凯斯——发现锂的电化学活性出奇的高。

The preservation, use, or processing of lithium is much more complex than other metals, resulting in this metal not being used for a considerable period of time after its discovery. At that time, the fate of lithium seemed to be forever sealed in the laboratory.

However, there are always scientists who are not reconciled. In 1913, two American chemical physicists, Gilbert Lewis and Frederick Keyes, discovered that lithium is surprisingly electrochemically active.

为此，他们进一步设计了经典的"三电极实验"，精确地计算出了锂的电极电势，并且大胆预言——锂是具有最低电位的电极材料。

这两位著名科学家的此言一出，便指引了无数科研工作者开展"将金属锂作为最终负极"的研究。理论提出来了，但继续往下一步的实践，就不那么顺利了。

Therefore, they further designed the classic "three-electrode experiment" to accurately calculate the electrode potential of lithium, and boldly predicted that lithium is the electrode material with the lowest potential.

As soon as the two famous scientists made the prediction, countless researchers began to study using metal lithium as the final negative electrode. Following the theory proposed, the practice went not well.

锂这个"顽童"太过于活泼，少有溶液是不与它发生反应的。所以，找到一种能与锂和谐相处的电解液也就成了实验的当务之急。

The "urchin" lithium is too reactive, and few solutions do not react with it. Therefore, finding an electrolyte that can live in harmony with lithium has become a priority in the experiment.

终于，在 1958 年，美国人威廉·哈里斯发现可以采用有机酯溶液作为锂金属原电池电解质，首次确定了金属锂与有机电解质的组合，为锂电池的发展奠定了基础。

Finally, in 1958, William Harris of the United States discovered that organic ester solutions could be used as electrolytes for lithium metal primary batteries, and for the first time determined the combination of metal lithium and organic electrolytes, laying the foundation for the development of lithium batteries.

人们对我越来越有兴趣，把各种电极材料和我放在一起做试验！

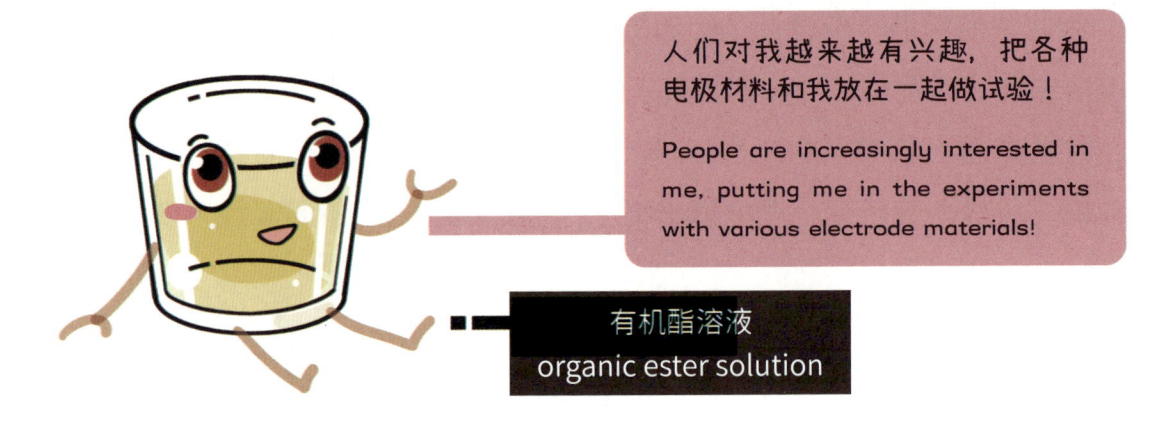

People are increasingly interested in me, putting me in the experiments with various electrode materials!

有机酯溶液
organic ester solution

哈里斯的重要发现，使得人们对金属锂应用于可充电电池的研究热情进一步高涨。在接下来的十多年里，人们有了至关重要的发现——固体电解质膜（SEI）。

SEI 的发现解决了锂应用于可充电电池的最大问题。此时，人们距离造出可充电锂电池只剩下一步之遥。

Harris's important discovery has further fueled people's enthusiasm for the application of metal lithium in rechargeable batteries. In the following decade, people made a crucial discovery—solid electrolyte interphase (SEI).

The discovery of SEI has solved the biggest problem of lithium application in rechargeable batteries. At this point, people are only one step away from producing rechargeable lithium batteries.

小贴士 Tips

SEI 是由金属锂和有机电解液反应产生的一层钝化膜，它附着在金属锂的表面，对锂起着稳定和保护的作用。

同时，SEI 就像传送带，能够来回传输电池中的工作物质——锂离子。

SEI is a passivation layer that forms as a result of the reaction between metal lithium and the organic electrolyte. This layer adheres to the surface of lithium, playing a crucial role in stabilizing and protecting the metal.

Meanwhile, SEI is like a conveyor belt that can transport the working substance, lithium-ion, back and forth in the battery.

20世纪70年代，在全球石油危机的背景下，美国为了减少对石油进口的过度依赖，开始大力发展新能源和储能技术，对锂电池的研究与开发也就成为重中之重。

美国某石油公司专门建立了可充电锂电池研究实验室，招揽了大批物理界和化学界的顶级人才，其中就包括英国科学家、2019年诺贝尔奖化学奖获得者斯坦利·惠廷厄姆。

In the 1970s, in the context of the global oil crisis, the United States began to develop new energy and energy storage technologies in order to reduce its over-reliance on oil imports, and the research and development of lithium batteries became a top priority.

An American oil company established a research laboratory dedicated to rechargeable lithium batteries, attracting a large number of top talents from the fields of physics and chemistry, including British scientist Stanley Whittingham, who was awarded the Nobel Prize in Chemistry in 2019.

经过近五年的研究后，惠廷厄姆以及他的同事以二硫化钛作为正极材料、锂铝合金作为负极材料，制成了世界上第一块可充电的锂电池，这种电池已经十分接近今天的锂离子电池。

After nearly five years of research, Whittingham and his colleagues developed the world's first rechargeable lithium battery, using titanium disulfide as the positive electrode material and lithium-aluminum alloy as the negative electrode material. This battery was very close to today's lithium-ion batteries.

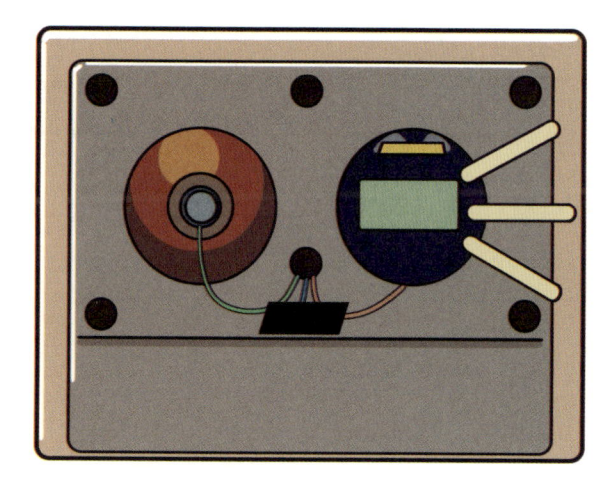

世界上第一块可充电的锂电池

The world's first rechargeable lithium battery

锂电池为人们的生活带来了便利，但问题也随之而来。

一是锂电池无法量产的问题。作为世间上最活泼的元素之一，金属锂在常温下就能与氮气发生反应，如果组装过程稍有不慎，泄入了空气，轻则电池报废，重则起火爆炸。

Lithium batteries have brought convenience to people's lives, but problems have also arisen.

The first problem is that this battery cannot be mass-produced. As one of the most active elements in the world, lithium metal can react with nitrogen at room temperature. If the assembly process is slightly careless and leaks air, the battery will be scrapped at best, and it will catch fire and explode at worst.

二是安全问题。以金属锂作为电极材料的电池，在早期的使用过程中有严重的安全隐患。惠廷厄姆带领团队开发的锂电池问世不到半年，就因多起起火爆炸事故被召回。

The second problem is safety. Batteries that use metal lithium as an electrode material have serious safety hazards in the early use process. The lithium battery developed by Whittingham's team was recalled due to multiple fires and explosions less than half a year after it came out.

电池小知识 Battery tip

金属锂会在负极上结晶，形成树枝状的金属锂——锂枝晶。当锂枝晶累积到一定程度，便会刺破隔膜，造成锂电池内部短路，引发电池自燃，甚至爆炸。

Metal lithium can crystallize on the negative electrode, forming lithium dendrites. When these lithium dendrites grow to a certain extent, they can puncture the separator, causing a short circuit inside the lithium battery, which can lead to self-ignition or even explosion.

这种情况一直持续到1987年，加拿大某能源公司，推出了用二氧化钼作为正极、金属锂作为负极的锂电池，这种锂电池受到了全世界的追捧，成为一款革命性产品，也是第一款真正意义上广泛商业化的锂电池。

This situation continued until 1987 when a Canadian company, launched a lithium battery that used molybdenum dioxide as the positive electrode and metallic lithium as the negative electrode. This lithium battery was highly sought after worldwide, becoming a revolutionary product and the first commercial lithium battery.

"锂"想的破灭与重生
The Rise and Fall of "Lithium" Battery

即便实现了量产,那就能说明锂电池安全了吗?这还需要经过时间的检验。

让人没想到的是,意外很快降临。1989 年,上述加拿大能源公司的锂电池产品发生了爆炸事故,这引发了整个市场对锂电池的恐慌。他们不得不召回了所有产品,然后在当年年底就宣布破产。

此次事件后,日本某电子公司宣布永久放弃把金属锂作为负极用于可充电电池的技术路线。

Even if mass production is achieved, are lithium batteries safe? It takes time to see.

Surprisingly, an unexpected event occurred soon. In 1989, an explosion of lithium battery from the Canadain company triggered panic in the entire market. They had to recall all their products and declared bankruptcy at the end of that year.

After this incident, a Japanese electronics company announced its permanent abandonment of the technology route of using metallic lithium as a negative electrode for rechargeable batteries.

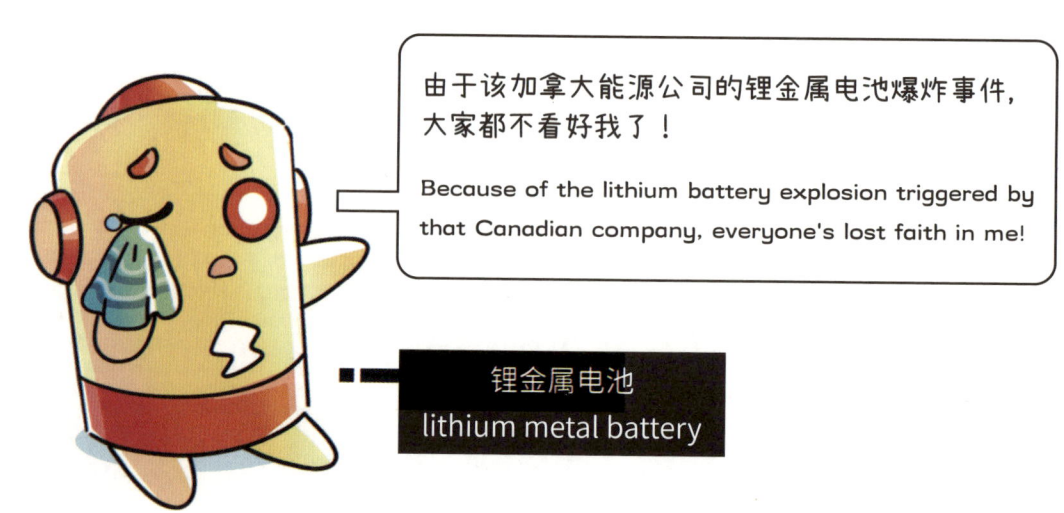

由于该加拿大能源公司的锂金属电池爆炸事件,大家都不看好我了!

Because of the lithium battery explosion triggered by that Canadian company, everyone's lost faith in me!

锂金属电池
lithium metal battery

由于安全性差，锂金属电池很快便从大众视野中消失了，但人们对锂电池的研究并没有因此而终止。为了让锂电池变得更加安全，有研究工作者意识到——必须从负极材料的替代入手。

然而，负极材料的更换也会面临一系列问题：锂金属的电势很低，使用其他的化合物做负极就一定会提高负极电势，而这样一来锂电池整体的电势差就会减小，电池能量密度就会下降。

因此，除了寻找对应的高压正极材料以外，还需要找到合适的电池的电解液，以匹配正负极电压以及保证循环稳定，同时电解液的电导率和耐热性能还要好。

Due to poor safety, lithium metal batteries quickly disappeared from the public, but research on lithium batteries did not cease. To make lithium batteries safer, researchers realized that it was essential to start with the replacement of the negative electrode materials.

However, replacing them also presents a series of challenges: lithium metal has a low potential, and using other compounds as the negative electrode will inevitably increase the negative electrode potential. Consequently, this will reduce the overall potential difference of the lithium battery, leading to a decrease in battery energy density.

Therefore, in addition to finding corresponding high-voltage positive electrode materials, it is also necessary to identify suitable electrolytes for the battery to match the voltages of electrodes while ensuring stable cycling. Additionally, the electrolyte should have good conductivity and thermal stability.

换掉锂金属容易，但是高压的正极材料和合适的电解液去哪里找呢？

Replacing lithium is easy, but where to find high-voltage positive electrode materials and suitable electrolytes?

被这些问题困扰了很久后，研究者们才找到了一个较为满意的方案，那就是寻找一种嵌入化合物，来代替金属锂作为负极。

直到1972年，法国科学家——米歇尔·阿芒迪提出，可以用一种电位较低的嵌锂插层化合物，来代替安全性较低的金属锂作为负极，并保证锂离子在正负极之间实现可逆嵌入与脱出。

这个理论被称为"摇椅式"概念，一直被沿用至今。这一概念阐明了锂离子电池的基本工作原理，将锂离子在正负极之间的穿梭形象地比作摇椅的摇动。

After being troubled by these issues for a long time, researchers eventually found a more satisfactory solution: to replace metallic lithium with an embedded compound as the negative electrode.

In 1972, French scientist Michel Armand proposed that a low-potential lithium intercalation compound could be used as the negative electrode to replace the less safe metallic lithium, ensuring the reversible insertion and extraction of lithium-ions between two electrodes.

This theory is known as the "rocking-chair" concept, which has stood the test of time. It explains how lithium-ion batteries work, comparing the shuttling lithium-ions between electrodes to the rocking of chairs.

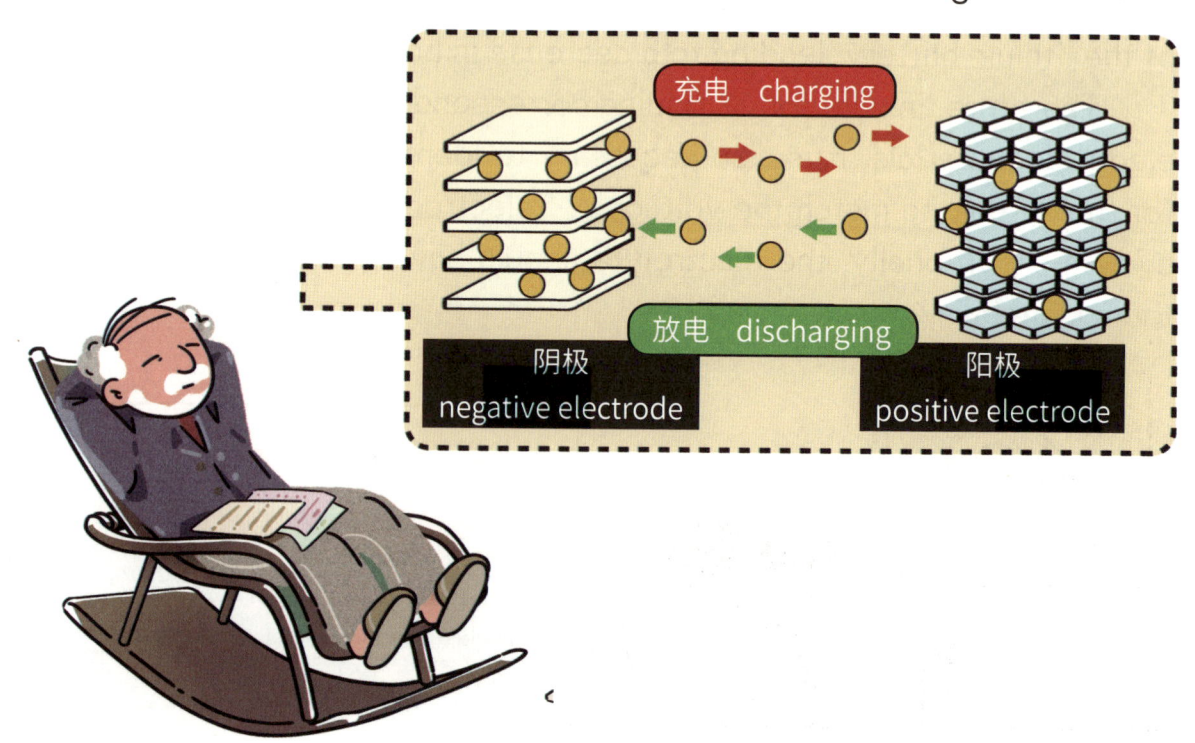

1980 年，美国科学家约翰·古迪纳夫受惠廷厄姆的启发，将正极材料的选择范围由金属硫化物调整为金属氧化物，这样既保证了正极材料在高电位时的稳定性，又提高了全电池的电压。

钴酸锂、锰酸锂以及磷酸铁锂三种正极材料，均出自古迪纳夫之手。于是，古迪纳夫的三次飞跃式的研究突破让锂离子电池迎来了曙光。

In 1980, inspired by Whittingham, American scientist John Goodenough expanded the selection range of positive electrode materials from metal sulfides to metal oxides. This change not only ensured the stability of positive electrode materials at high potentials but also increased the overall voltage of the battery.

These three positive electrode materials including lithium cobalt oxide, lithium manganese oxide, and lithium iron phosphate were all developed by Goodenough. As a result, Goodenough's three groundbreaking researches brought new hope to lithium-ion batteries.

虽然我没从价值 350 亿的锂电池市场中赚到一分钱，但是我不后悔，这是我应该做的事情。

Although I didn't make a penny from the lithium battery market worth 35 billion, I have no regrets. This is what I was meant to do.

古迪纳夫
Goodenough

锂离子电池的诞生
The Birth of Lithium-ion Battery

正极的问题基本解决后，攻克负极这一难题成为重中之重。1985 年，日本科学家吉野彰采用石油焦为负极，并以钴酸锂作为正极，开发出了世界上第一个锂离子电池。

1991 年，日本某公司对全新的锂离子电池进行商业化生产，极大地推动了锂离子电池及相关领域的发展。

锂离子电池的问世为人类打开了安全、可充电世界的大门。作为可移动便携能源，锂离子电池屡屡大展身手，从通讯、办公到出行，到处都有它的影子。

With the issues related to the positive electrode largely resolved, tackling the challenges of the negative electrode became a top priority. In 1985, Japanese scientist Akira Yoshino developed the world's first lithium-ion battery using petroleum coke as the negative electrode and lithium cobalt oxide as the other.

In 1991, a company in Japan began the commercial production of the new lithium-ion battery, which greatly propelled the development of lithium-ion batteries and related fields.

The advent of lithium-ion battery opened the door to a safe and rechargeable world for humanity. As a portable energy source, lithium-ion batteries have showcased their capabilities in various applications, from communication and office work to transportation, making their presence felt everywhere.

为了表彰科学家们在锂电池的研究开发过程中作出的卓越贡献，2019 年，瑞典皇家科学院宣布，将 2019 年诺贝尔奖化学奖颁发给在锂离子电池的发展史当中最具有影响力的 3 位学者，他们分别是惠廷厄姆、古迪纳夫和吉野彰。

To honor the outstanding contributions of scientists in the research and development of lithium battery, the Royal Swedish Academy of Sciences announced in 2019 that the Nobel Prize in Chemistry for that year would be awarded to three scholars who have made the most significant impact in the history of lithium-ion battery development. They are Goodenough, Whittingham, and Akira Yoshino.

这三位都是我的"父亲"！
These three are all my "fathers"!

锂离子电池的未来
The Future of Lithium-ion Battery

从 20 世纪末到 21 世纪初，几乎所有新出现的机器都由锂电池驱动。在未来，随着可穿戴设备的广泛运用，锂离子电池的应用范围将会进一步扩大。

From the late 20th century to the early 21st century, almost all emerging technological devices have been powered by lithium batteries. In the future, with the widespread wearable devices, the application range of lithium-ion battery will further expand.

目前，中国是全世界最大的锂电池生产国，也是最大的出口国。我国的锂电池产业正迎来飞速发展期，锂离子电池实现了从"中国制造"到"中国智造"的大转变。

Currently, China is the largest producer and exporter of lithium batteries in the world. The lithium-ion battery industry is experiencing rapid development, marking a significant transition from "Made in China" to "Intelligent Manufacturing in China".

与此同时，我国在锂电池产业链上的研发与扩产仍在继续，锂离子电池领域也因此出现了更多创新的方向。例如，全固态锂电池、锂硫电池等。据称，全固态锂电池能使锂离子电池的安全性大幅度提升，而锂硫电池的能量密度或可达现有锂离子电池的 2 倍，这些都是令人振奋的消息！

创新仍在继续，随着锂电池技术的不断迭代升级，角逐永不停止……

Meanwhile, China's ongoing research and development, along with expansion in the lithium battery industry chain, are driving further innovations in lithium-ion batteries. For instance, advancements include all-solid-state lithium batteries and lithium-sulfur batteries. Reportedly, all-solid-state lithium batteries significantly enhance the safety of lithium-ion batteries. The energy density of lithium-sulfur batteries potentially doubles that of current lithium-ion batteries. These are truly exciting!

Innovation continues to drive the upgrading technology, and the competition never stops…

话说燃料电池
The Story of Fuel Battery

自从电被人类发现并投入生活和工业使用，如何低成本且大规模发电就成了几代科学家研究的重点。燃料电池作为一种发电效率高、环境污染小但成本也较高的发电装置，成为多年来研究的一大难点。

燃料电池是一种把燃料所具有的化学能直接转换成电能的化学装置，又称"电化学发电器"。它是继水力发电、热能发电和原子能发电之后的第四种发电技术。

Since electricity was discovered and integrated into daily life and industrial use, generating electricity at low cost and on a large scale has been a focal point of research for generations of scientists. Fuel battery, as a highly efficient and environmentally friendly power generation device with high costs, has become a major challenge in research for many years.

Fuel battery is a chemical device that directly converts the chemical energy of fuel into electrical energy, also known as "the electrochemical generator". It represents the fourth power generation technology, following hydropower, thermal power, and atomic power.

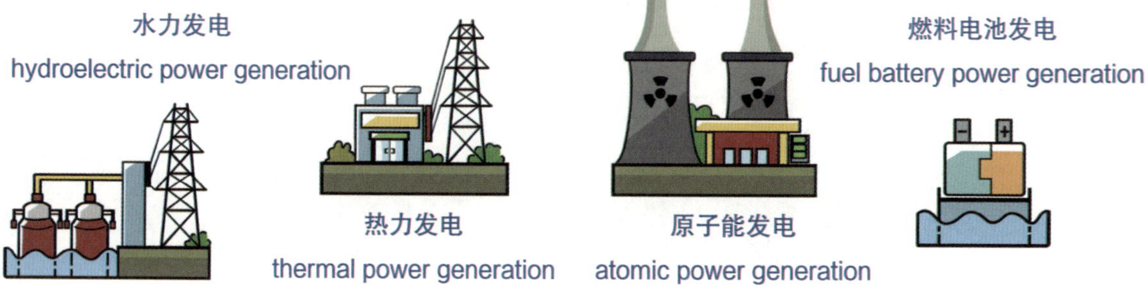

水力发电
hydroelectric power generation

热力发电
thermal power generation

原子能发电
atomic power generation

燃料电池发电
fuel battery power generation

燃料电池的起源
The Origin of Fuel Battery

燃料电池已经有 180 余年的历史。世界上第一块燃料电池的产生，建立在人类对氢气的认知上。

早在 1776 年，英国科学家亨利·卡文迪许便在金属锌与盐酸反应后捕获到了氢气，于是他认定：氢，是一种独特的元素。

Fuel battery boasts a history of over 180 years. The production of the world's first fuel battery stemmed from humanity's knowledge of hydrogen gas.

Back in 1776, British scientist Henry Cavendish captured hydrogen gas following the reaction between zinc metal and hydrochloric acid. Thus, he concluded that hydrogen was a unique element.

后来，德国化学家克里斯蒂安·尚班在 1838 年首次提出了燃料电池的原理，并将其发表在了当时著名的科学杂志上。

Subsequently, in 1838, German chemist Christian Schönbein first introduced the principle of fuel battery and published it in a renowned scientific journal of the time.

1840年前后，英国物理学家威廉·格罗夫受到尚班的理论的启发，在水电解研究中首次发现了氢气可以用来发电的现象，并制造了世界上第一台氢氧燃料电池。

Around 1840, inspired by Schönbein's theory, British physicist William Grove discovered in his research on water electrolysis that hydrogen could be harnessed to generate electricity. He also made the world's first hydrogen-oxygen fuel battery.

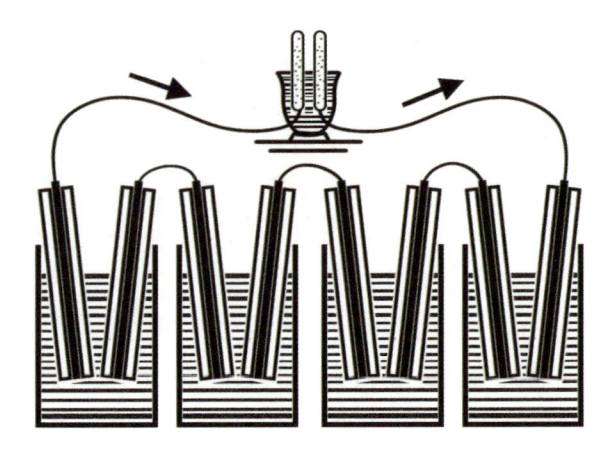

格罗夫初代燃料电池草图
sketch of Grove's first-generation fuel battery

图片引自《电池工业》2000年12月第5卷第6期《燃料电池的历史、现状和未来》图1

Quoted from *Battery Industry*, Volume 5, Issue 6, December 2000, *History, Current Status, and Future of Fuel Battery* Figure 1

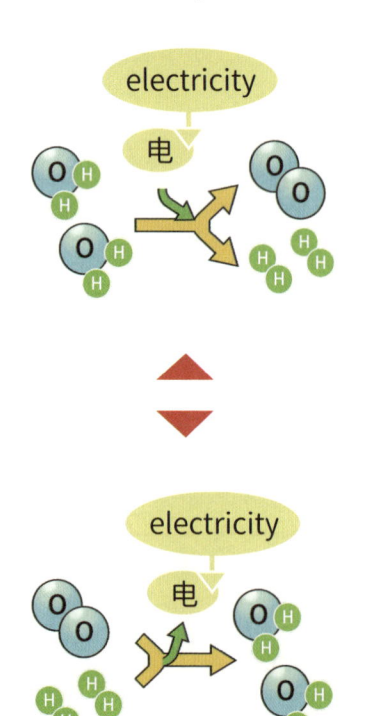

原理：水被电解为氢气和氧气，其逆反过程为氢气在铂的催化作用下变为氢离子和氧气发生化学反应，产生水和电。

大意就是氢气在铂的催化作用下生成氢离子，氢离子通过电解液传输到氧气侧生成水，电子通过外电路传输发电。

Principle: water is electrolyzed into hydrogen and oxygen. In the reverse process, hydrogen, catalyzed by platinum, transforms into hydrogen-ions and reacts with oxygen to produce water and electricity.

The general idea is that hydrogen generates hydrogen ions under the catalysis of platinum, which are transported to the oxygen side through the electrolyte to generate water, and electrons are transmitted through an external circuit to generate electricity.

因此，葛洛夫被称为"燃料电池之父"。

Hence, Grove is widely recognized as the "father of fuel battery".

不过，由于燃料电池在理论以及材料方面的不完善，伴随着火力发电和蒸汽发电技术逐渐成熟并开始被大规模投入使用，几相对比下，价格昂贵的燃料电池显得毫无竞争力，只能被退回到实验室继续研究。

However, the theory and materials of fuel battery are not yet fully developed. As thermal power generation and steam power generation technologies gradually mature and begin to be put into use on a large scale, the expensive fuel batteries appear uncompetitive and can only be returned to the laboratory for further research.

一个燃料电池成本的 60% 是催化剂——铂，太贵了！

60% of the cost of a fuel battery is from platinum, the catalyst, which is too expensive!

小贴士 Tips

铂就是铂金，它是一种化学元素，俗称"白金"。

Platinum is a chemical element commonly known as the "white gold" in Chinese (*baijin*).

虽然成本较高，但燃料电池发电非常高效、简单，而且比当时盛行的火力发电也可靠许多。所以那时候的科研工作者们一直都没有减少对它的期待。

Although the cost is high, fuel battery is very efficient, simple, and much more reliable than the prevalent thermal power in power generation at that time. So, researchers never gave up their expectations for it.

> 燃料电池能够直接将化学能转化为电能，能量转换效率可达45%～60%，比火力发电以及核能发电都要高得多。
>
> Fuel battery can directly convert chemical energy into electrical one, with the conversion efficiency of 45%–60%, much higher than thermal and nuclear power generation.

直到将近 100 年后的 1932 年，英国发明家弗朗西斯·培根制造出第一个可以投入实际生产的燃料电池。这是第一个实际意义上的碱性燃料电池（AFC），因而又被称为"培根碱性电池"。

Till almost 100 years later in 1932, British inventor Francis Bacon created the first fuel battery that could be put into actual production. This is the first substantial alkaline fuel battery (AFC), hence also known as the "Bacon alkaline battery".

> 定个小目标，先满足一台电焊机所需的动力！
>
> Let's start small—power an electric welding machine!

open-circuit voltage converter

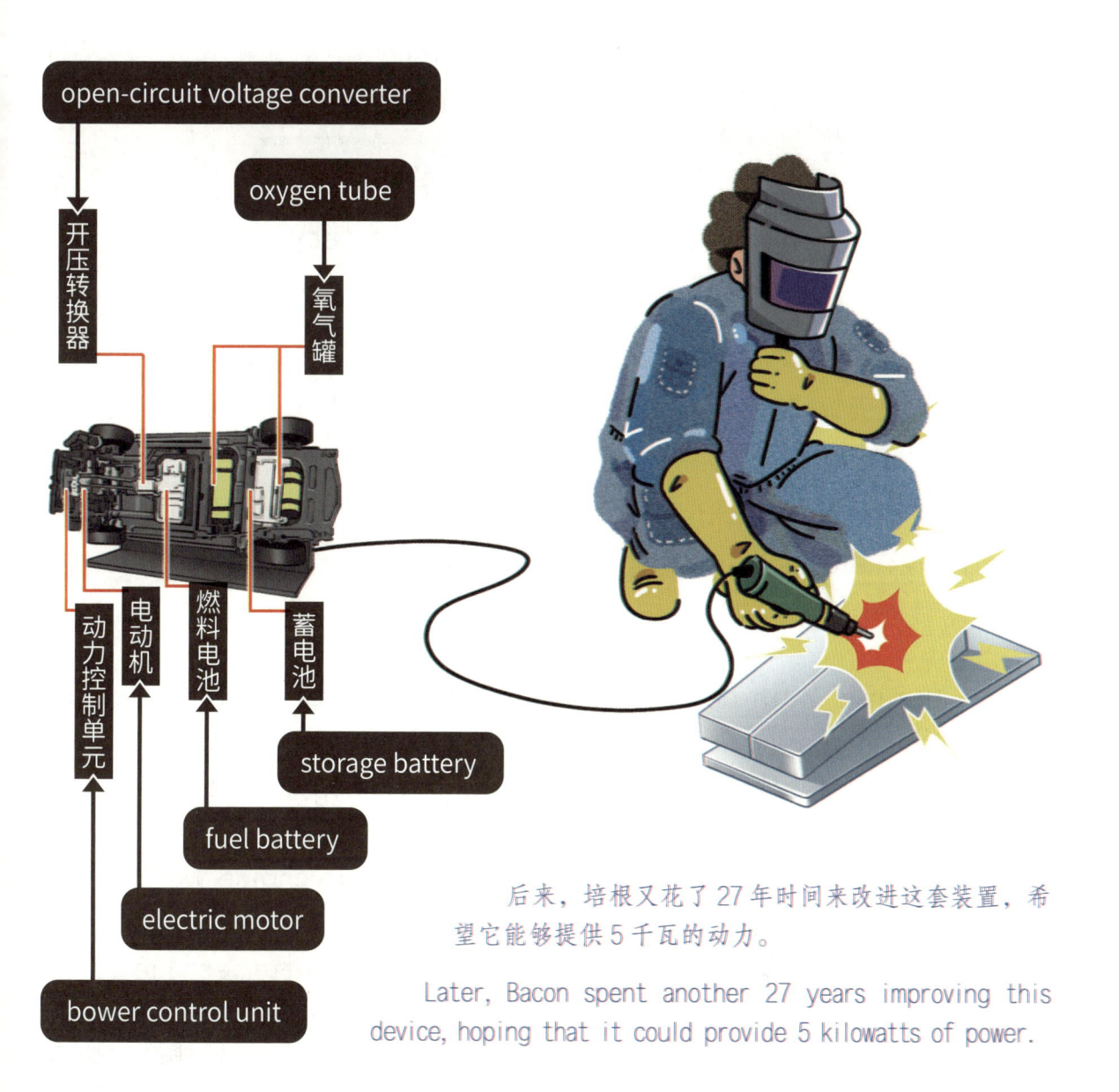

开压转换器

oxygen tube

氧气罐

动力控制单元

电动机

燃料电池

蓄电池

storage battery

fuel battery

electric motor

bower control unit

后来，培根又花了27年时间来改进这套装置，希望它能够提供5千瓦的动力。

Later, Bacon spent another 27 years improving this device, hoping that it could provide 5 kilowatts of power.

燃料电池的发展
Development of Fuel Battery

20世纪中期，人类对探索太空的需求成为燃料电池技术发展的最大推动力。尤其在载人航天领域，干电池太重，太阳能当时太贵，核能又太危险，现有的能源几乎都被排除了，除了研发一种新的能源以外别无他法。

终于，燃料电池因体积小、容量大的优势，在众多电池中脱颖而出，并且在1960年经过开发和应用新型催化剂后被成功升级，寿命变得更长，效率也更高了。

In the mid-20th century, humans' demand for exploring space became the biggest driving force for the development of fuel battery technology. Especially in manned spaceflight, when almost all existing energy options were ruled out including heavy dry batteries, then-costly solar energy, and then-dangerous nuclear energy, no choice was left but to develop a new energy source.

Finally, fuel battery stood out among many batteries due to its advantages of small size and large capacity, and was successfully upgraded after the development and application of new catalysts in 1960, resulting in longer lifespan and higher efficiency.

20 世纪 60 年代，燃料电池首次被用在"阿波罗"登月飞船上，作为太空计划中电力和水的来源。后来，燃料电池多次被用于航天飞行，为人类探索太空作出了卓越贡献。那么，燃料电池是如何从太空中下到地面的呢？

进入 20 世纪 70 年代后，随着技术的不断进步，氢燃料电池逐步被运用于发电和汽车。

In the 1960s, fuel batteries were first used on the *Apollo* lunar module as a source of electricity and water for the space program. Later, fuel batteries were used in spaceflight many times, making outstanding contributions to human exploration of space. So, how did fuel batteries transfer from space to the earth?

In 1970s, with the continuous advancement of technology, hydrogen fuel batteries were gradually applied to power generation and automobiles.

全球第一款燃料电池汽车——Electrovan

the world's first fuel battery vehicle—Electrovan

1966 年，美国某汽车公司推出了全世界第一款燃料电池汽车，它的动力系统由 32 个串联的薄电极燃料电池模块组成，持续输出功率为 32 千瓦，峰值功率达到 160 千瓦，完美展示了燃料电池技术的可行性潜力。

In 1966, an American car company introduced the world's first fuel battery vehicle. Its power system consisted of 32 series-connected thin electrode fuel battery modules, with a continuous output power of 32 kilowatts and a peak power of 160 kilowatts, perfectly demonstrating the potential of fuel battery technology.

目前，燃料电池汽车加一次氢气，最长可续航 500 千米以上，但价格远远低于使用燃油汽车。

At present, fuel battery vehicles can have a range of over 500 kilometers with just one hydrogen injection, and the price is much lower than that of gasoline vehicles.

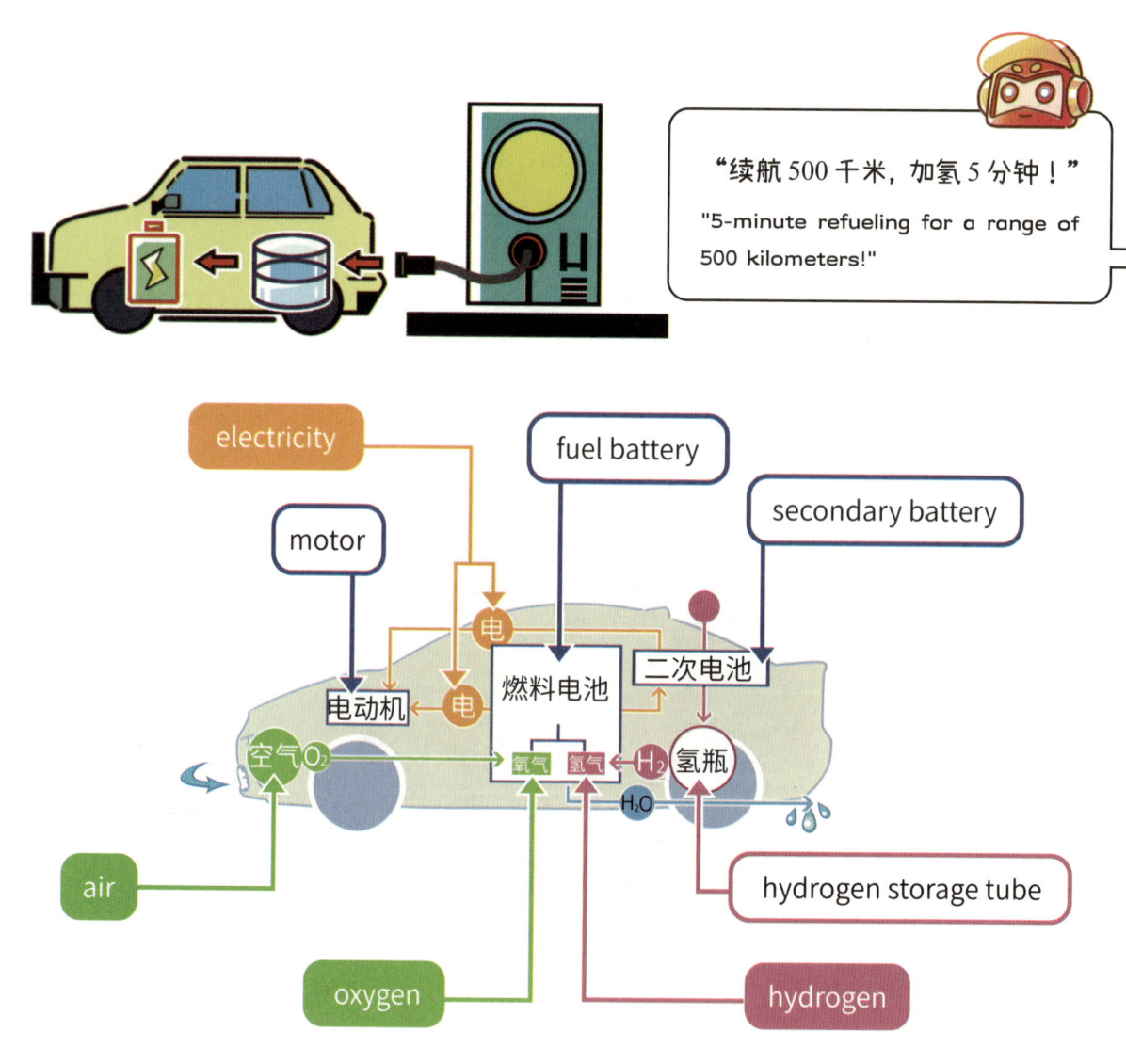

"续航 500 千米，加氢 5 分钟！"

"5-minute refueling for a range of 500 kilometers!"

汽车中燃料电池的工作原理

the working principle of fuel battery in automobiles

研究燃料电池的必要性
The Necessity of Researching Fuel Battery

氢能源是地球上储量最大的能源之一。当氢气与氧气在燃料电池中发生反应时，只会产生热量和水，是最清洁的能源。

Hydrogen energy is one of the largest energy sources on earth. When hydrogen and oxygen react in a fuel battery, only heat and water are produced, making it the cleanest energy source.

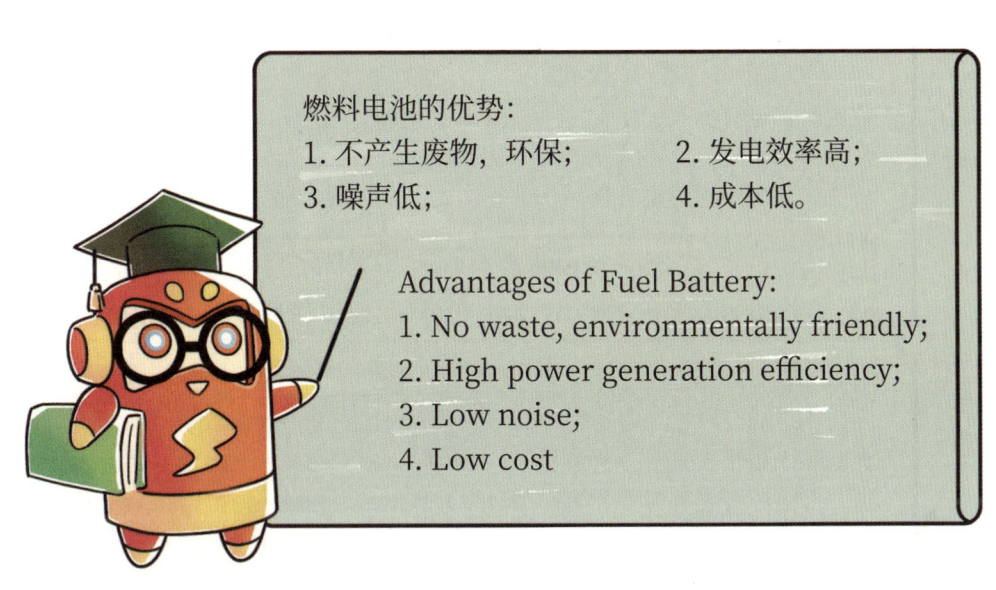

燃料电池的优势：
1. 不产生废物，环保；　2. 发电效率高；
3. 噪声低；　　　　　　4. 成本低。

Advantages of Fuel Battery:
1. No waste, environmentally friendly;
2. High power generation efficiency;
3. Low noise;
4. Low cost

随着时代的发展，燃料电池技术逐渐成熟，性能不断提升，应用也在不断拓展，可以说是已经取得了重大进展。目前，已经发展到了第四代燃料电池技术，并在一定程度上实现了商业化。

With the development of the times, fuel battery technology has gradually matured. It can be said that significant progress has been made in its improved performance and expanded applications. At present, the fourth-generation fuel battery technology has been developed and commercialized to a certain extent.

Generation 1: 磷酸型燃料电池（PAFC）技术 / phosphoric acid fuel battery (PAFC) technology

目前较为成熟的应用技术之一，已经成功进入商业化应用和批量生产。但是由于其成本太高，目前主要用作区域性电站的供电、供热。

As one of the mature technologies to date, it has been commercially used and mass-produced. However, due to its high cost, it is mainly used for power supply and heating in regional power stations.

Generation 2: 熔融碳酸盐燃料电池（MCFC）技术 / molten carbonate fuel battery (MCFC) technology

目前主要应用于设备发电。

It is primarily applied in power generation for equipment.

Generation 3: 固体氧化物燃料电池（SOFC）技术 / solid oxide fuel battery (SOFC) technology

因其全固态结构、更高的能量效率的特点，对煤气、天然气、混合气体等多种燃料气体具有广泛的适应性。

Due to the all-solid-state structure and higher energy efficiency, it exhibits a broad adaptability to various fuel gases, including coal gas, natural gas, and mixed gas.

Generation 4: 质子交换膜燃料电池（PEMFC）技术 / proton exchange membrane fuel battery (PEMFC) technology

具有较高的能量效率和能量密度，体积重量小，更安全可靠，正逐渐拓展其商业应用。

It is known for its high energy efficiency and energy density, compact size, and greater safety and reliability. It's now gradually expanding its commercial applications.

燃料电池技术的研究与开发是21世纪重要的高科技产业之一，燃料电池已被应用于汽车工业、能源发电、船舶工业、航空航天、家用电源等行业，受到全世界的高度重视。

The research and development of fuel battery technology is one of the important high-tech industries in the 21st century. Fuel batteries have been used in the automotive industry, energy power generation, shipbuilding industry, aerospace, household power supply, and other industries, attracting worldwide attention.

话说碱锰电池
The Story of Alkaline Manganese Battery

　　碱锰电池的全称为"碱性锌锰电池"，它是由锌锰干电池发展而来的，碱锰电池可以说是锌锰干电池的升级换代版高性能电池。

　　锌锰干电池是历史最悠久的干电池，自100多年前诞生以来，它的发展经历了漫长的演变。

The full name of alkaline manganese battery is "alkaline zinc-manganese battery", which is developed from zinc-manganese dry battery. It can be said to be an upgraded version of high-performance battery of zinc-manganese dry battery.

Zinc-manganese dioxide dry battery is the oldest type of dry batteries around. Since their birth over 100 years ago, they've gone through a lengthy journey of evolution.

限制电池汞含量的工作正在分步骤实施，首先实现低汞，最终达到无汞。你，快被开除了！

Efforts to limit the mercury content in batteries are being rolled out step by step, starting with the aim of low-mercury levels and eventually reaching mercury-free status. You're about to be the history!

呜呜呜……怎么办……
Aww…what should I do?…

碱锰电池与锌锰干电池最大的不同在于"碱性"与"酸性"的区别。且锌锰干电池含汞，不可以随生活垃圾处理，需要单独回收，随意丢弃会造成环境污染；而碱锰电池是不含汞的，不需要单独回收。另外，锌锰干电池相较于碱锰电池来说，虽然体积小、携带更加方便，但放电功率比较低，低温性能比较差；而碱锰电池即便是在低温环境下也可以正常工作，因此在高寒地区，碱锰电池更适用。此外，碱锰电池的电极结构与普通干电池相反，增大了正、负极之间的相对面积，使电池性能更加稳定且容量更大。目前，碱锰电池正在慢慢取代酸性锌锰电池而成为主流。

　　The biggest difference between alkaline manganese battery and zinc-manganese dry battery is "being alkaline" or "being acidic". And zinc-manganese dry batteries contain mercury and cannot be disposed of with domestic waste. They need to be recycled separately, and discarding them at will cause environmental pollution; while alkaline manganese batteries do not contain mercury and do not need to be recycled separately. Additionally, zinc manganese dry batteries are smaller and more portable, but they have lower discharge power and poorer performance in low temperatures. While alkaline manganese batteries can still work in cold places, making them a better choice for freezing climates. What's more, alkaline manganese batteries also have a different electrode structure compared to regular dry batteries. This design increases the relative areas between the positive and negative terminals, making the battery more stable and giving it a larger capacity. Today, alkaline manganese batteries are slowly replacing zinc-manganese dry batteries and becoming the main type of battery used.

话说镍镉电池
The Story of Nickel-cadmium Battery

19 世纪 90 年代，有个叫恩斯特·容纳的年轻人做了一件了不起的事情。

容纳所在的瑞典，由于地处北欧，日照时间短，冬季长达半年，每年 12 月有很多地方几乎整日不见阳光。

In the 1890s, a young man named Ernst Jungner did something amazing.

Jungner lived in Sweden, a country in Northern Europe. Because Sweden is so far north, the sunshine duration is short, and the winter lasts for half a year. Every December, many places there see almost no sunlight all day.

在没有电灯的年代，瑞典所有的家庭都备有大量蜡烛。那时，蜡烛就像食物和水一样，是人们生活的必需品，但这也带来了火灾隐患。

Back then, there were no electric lights. Candles, a must-have household item in Sweden, were as important as food and water. But this also brought about a fire hazard.

于是，年仅 19 岁的容纳便设计了一个由热电偶串联组成的火灾报警系统，用来提示火情。

Therefore, at the age of 19, Jungner designed a fire alarm system composed of thermocouples connected in series to indicate a fire.

原则上这个报警器可以很好地工作，但是由于它的电力来源于当时性能还很不稳定的干电池，导致报警器有些时候并不能很好地发挥作用。这让容纳大为头疼，他下定决心要创造一种性能优于干电池和铅蓄电池的电源，并展开了艰苦的实验。

终于，在1899年，容纳的目标得以实现，他发明出最早的镍镉电池，也是最早出现的干式充电电池，这种电池解决了早期的铅酸蓄电池漏液的问题。

经过数十年的改良，1930年的镍镉电池已经可以承受大电流密度的放电，开始被广泛用于军事领域。它在第二次世界大战中大展风采，出现在通信电台、坦克、装甲车辆、飞机等装置的配套电源里。

In principle, this alarm could work well, but because its power came from dry batteries that were still very unstable at the time, the alarm did not work very well at times. This was a big headache for Jungner. He determined to create a power supply that outperformed dry batteries and lead-acid batteries, and carried out painstaking experiments.

Finally, in 1899, Jungner achieved his goal by inventing the earliest nickel-cadmium battery and also the first dry rechargeable battery. This battery solved the leakage problem of early lead-acid batteries.

After decades of refinement, the nickel-cadmium battery of 1930 was able to withstand high current density discharge and began to be widely used in the military field. It flourished in World War II, supporting power supplies in communication stations, tanks, armored vehicles, aircraft and other devices.

到了20世纪60年代，由于战争的缘故，镍镉电池在短期内得到迅速发展。由于可以满足高负载、大功率的需要，它还被用于卫星，在人类航天事业的发展中起到重大的作用。还有火箭、导弹、电子计算机、助听器等，这些领域中都有镍镉电池的身影。那个时候，它处于蓬勃发展的态势。不过，镍镉电池在废弃后对环境危害较大，并且在充电过程中可能由于内部受损而发生事故，基于环境保护与健康等原因，在2005年左右逐渐被镍氢电池取代。镍氢电池对自然环境没有污染，没有记忆效应，在镍氢电池被创造出来之前，镍镉电池可以说是电池行业的老大。但是，自镍氢电池诞生以后，短时间内便占领了镍镉电池的市场，巨无霸电池老大的地位也就慢慢易主了。

到如今，大部分发达国家已建议禁止使用镍镉电池，市场上已经不多见镍镉电池的存在了。

By the 1960s, due to the war, nickel-cadmium batteries experienced rapid development in a short period. Because they could meet the demands of high loads and high power, they were also used in satellites, playing a significant role in the development of space exploration. Nickel-cadmium batteries were found in rockets, missiles, electronic computers, hearing aids, and more, thriving at that time. However, since nickel-cadmium batteries posed a great environmental hazard after being discarded, and accidents might occur due to internal damage during charging, they were gradually replaced by nickel-metal hydride batteries around 2005 for environmental protection and health reasons. Nickel-metal hydride batteries have no pollution to the natural environment and no memory effect. Before their creation, nickel-cadmium batteries could be said to be the leader in the battery industry. However, since the birth of nickel-metal hydride batteries, they have occupied the nickel-cadmium battery market in a short period of time, and the position of the Big Mac battery leader has gradually changed.

Up to now, most developed countries have recommended banning the use of nickel-cadmium batteries, and they are becoming increasingly rare in the market.

未来电池之争
The Race of Future Batteries

随着更加智能化的时代的到来，未来势必将会是电池供能的时代。

As we enter a more intelligent age, the future is undoubtedly one of battery-powered energy—there's no turning back. With the advent of a more intelligent era, the future will be an era of battery power, which is absolutely necessary.

未来的电池
battery in the future

移动技术的发展在很大程度上取决于电池技术的发展，在当前的科技背景下，电池的发展可以说是直接影响到人类社会现代文明的发展。

从电动汽车到工业级太阳能电厂，电池都将是更清洁、更高效的能源系统的关键所在。

Mobile technology largely depends on the advancement of battery technology. In the current technological context, the development of batteries can be said to have a direct impact on modern civilization in human society.

From electric vehicles to industrial-scale solar power plants, batteries will be the key to a cleaner and more efficient energy system.

锂离子电池轻便、性价比不错，而且可重复充电，与下一代商用的化学电池相比，可以提供更高的能量密度，自从问世以来就成为了为移动设备供电的主要方式。然而，即使已经拥有了堪称完美的电力来源，研究者们依然对新型电池保持着极高的热情与想象力。

例如，我国在1991年首创的以铝-空气-海水为能源的新型电池，它被命名为"海洋电池"。

Lithium-ion batteries are lightweight, cost-effective, and rechargeable, providing higher energy density than next-generation commercial chemical batteries. Since their inception, they have become the leading way to power mobile devices. However, even with the perfect power source, researchers still maintain a high degree of enthusiasm and imagination for new batteries.

For example, in 1991, China developed a new battery powered by aluminum, air, and seawater—an invention known as the "ocean battery".

海洋电池
the ocean battery

海洋电池可谓是海洋用电设施的能源新秀，它以稳定可靠、长效、无污染的优势替换掉之前的锌锰电池等一次电池，以及需要先充电、再给电的镍镉电池等二次电池。

研究者们在积极探索新一代电池技术的道路上，对电池材料的选取方面也在不断地寻求突破，例如从铅酸蓄电池到磷酸铁锂电池、三元锂电池、钠离子电池等。有的研究者则选择在电池结构上下功夫，研发出如改良的刀片电池和弹匣电池，甚至是颠覆传统的固态电池。

新一代电池的竞争日益激烈，角逐者中不乏建立在颇为新奇的概念上的。

The ocean battery can be described as the energy newcomer of marine power facilities. With stable, reliable, long-lasting, and pollution-free advantages, they replace the previous zinc-manganese batteries and other primary batteries, as well as secondary batteries like nickel-cadmium batteries that need to be charged first and then offer power supply.

In the active pursuit of a new-generation battery technology, researchers are continually seeking breakthroughs in the selection of battery materials, such as lead-acid, lithium-iron phosphate, ternary lithium, and sodium-ion. Some researchers choose to work on battery structure, like the improved blade, the magazine, and even the revolutionary solid-state.

The competition for new-generation batteries is becoming increasingly fierce, with contenders introducing quite novel concepts.

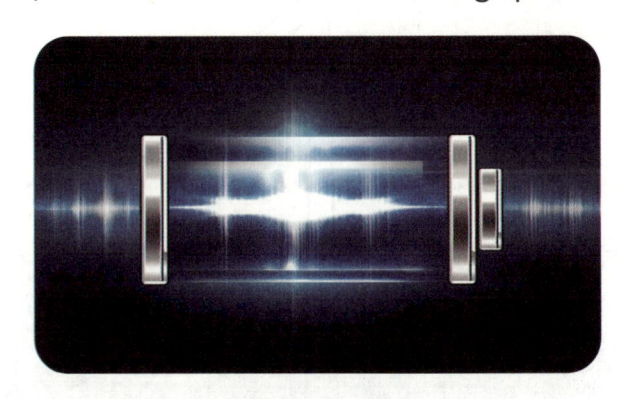

未来电池的构想
battery in the future

例如：液态电池；运行温度堪比汽车引擎工作温度的熔态金属电池；以盐水作为原料的电池；能将电池价格降低到现在的锂离子电池价格五分之一的锂-空气电池；还有能利用水和植物的光合作用来产生能量的植物电池；甚至还有比人的头发还小、储电量却很大，并且能够快速充电和放电的纳米电池等一系列神奇的"超级电池"。

　　机遇总是伴随着挑战，当前的人类就正在面临科学与技术前沿的挑战。尽管新型电池备受追捧，但离实现真正产业化尚有距离。从目前的发展情况来看，无论是镁电池、锌电池还是钠电池，在技术和材料等方面仍有诸多难题待解。

　　因此，在加快布局各种替代技术方案的同时，深挖锂电池性能潜力、提升产品质量仍是不二之选。

For example, there are liquid batteries, molten metal batteries operating at temperatures comparable to car engine working temperatures, batteries using saltwater as a raw material, lithium-air batteries that can reduce battery costs to one-fifth of the current lithium-ion battery prices, plant batteries capable of generating energy through water and plant photosynthesis, and even nanobatteries that are smaller than a human hair, yet store significant amounts of energy and enable rapid charging and discharging. These are all part of a remarkable lineup of "super batteries".

Opportunities always come with challenges, and humanity is currently facing challenges at the frontier of science and technology. Although new types of batteries are highly sought after, they are still far from achieving true industrialization. Based on current development trends, whether magnesium batteries, zinc batteries, or sodium batteries, many issues in technology and materials remain unresolved.

Therefore, while accelerating the deployment of various alternative technology solutions, it is still the best choice to explore the performance potential of lithium batteries and improve product quality.

图书在版编目（CIP）数据

电池简史：汉英对照 / 马建民主编；咪柯文化绘
图；李丝贝译. -- 成都：成都电子科大出版社，2025.
1. -- ISBN 978-7-5770-1487-6

Ⅰ．TM911-091

中国国家版本馆 CIP 数据核字第 2025KR1162 号

电池简史（中英对照版）
DIANCHI JIANSHI（ZHONG-YING DUIZHAO BAN）

马建民　主编　咪柯文化　绘　李丝贝　译

策划编辑	谢忠明　段　勇
责任编辑	赵倩莹
责任校对	蒋　伊
责任印制	段晓静

出版发行　电子科技大学出版社
　　　　　成都市一环路东一段 159 号电子信息产业大厦九楼　邮编 610051
主　　页　www.uestcp.com.cn
服务电话　028-83203399
邮购电话　028-83201495

印　　刷　成都久之印刷有限公司
成品尺寸　185 mm×260 mm
印　　张　8.5
字　　数　162 千字
版　　次　2025 年 1 月第 1 版
印　　次　2025 年 1 月第 1 次印刷
书　　号　ISBN 978-7-5770-1487-6
定　　价　66.00 元

版权所有，侵权必究